The Usborne
Complete Book
of the
Microscope

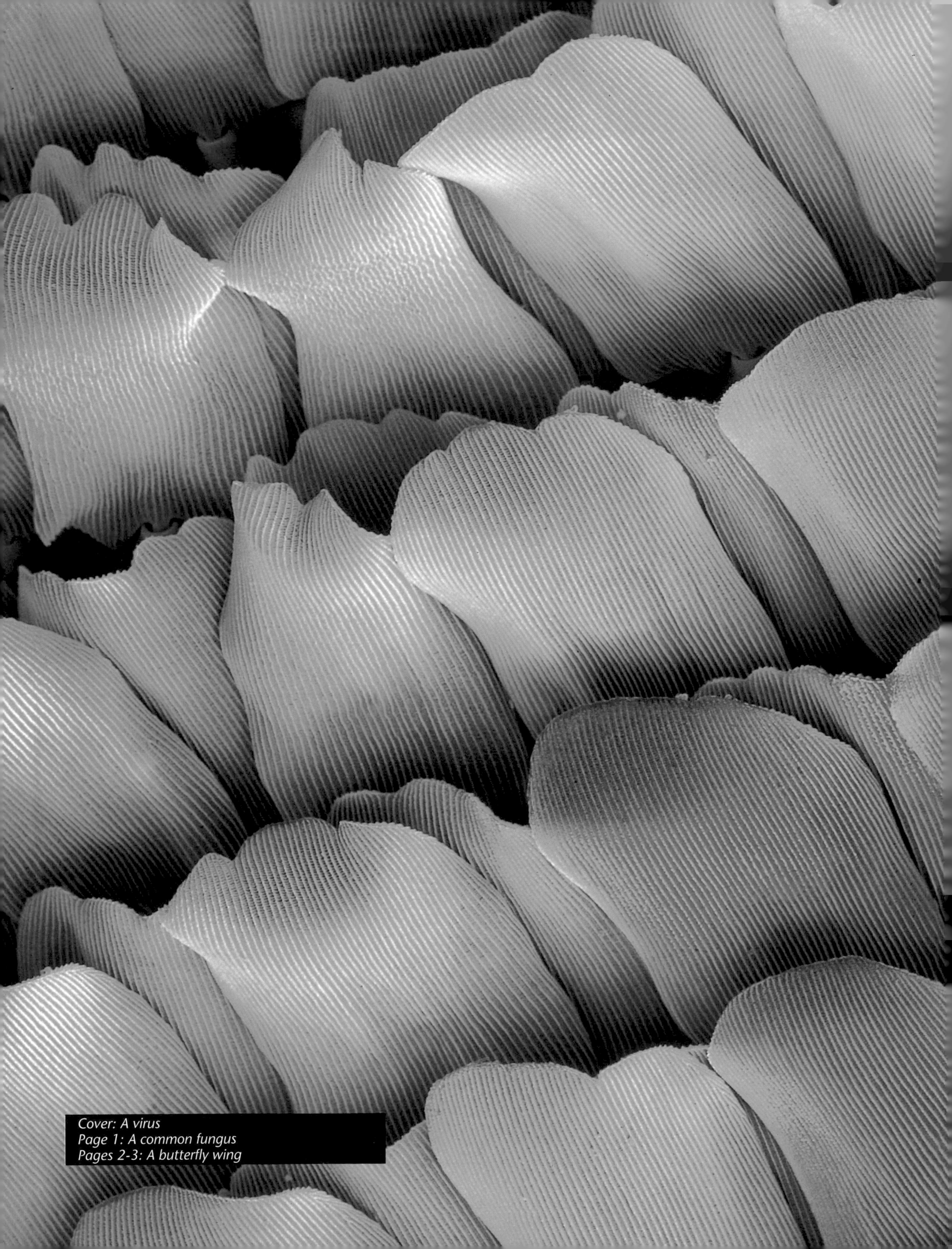

Cover: A virus
Page 1: A common fungus
Pages 2-3: A butterfly wing

The Usborne
Complete Book
of the
Microscope

Kirsteen Rogers

Edited by Paul Dowswell

Designed by Laura Fearn

Illustrated by Kim Lane, Gary Bines and Peter Bull

Additional design by Jane Rigby

Consultants: Revd. Professor Michael J. Reiss and Max Parsonage

Managing editor: Judy Tatchell

CONTENTS

Internet links

If you have access to the Internet, you can visit the websites recommended in this book. For links to these sites, go to the Usborne Quicklinks Website at **www.usborne-quicklinks.com** and enter the keyword "microscope". Please read and follow the Internet Safety Guidelines shown in Usborne Quicklinks.

The sites in this book have been selected by Usborne editors as suitable, in their opinion, for children, and the links in Usborne

Quicklinks are regularly reviewed and updated. However, the content of a website can change at any time and Usborne Publishing is not responsible for the availability or content of sites other than its own.

We recommend that children are supervised while on the Internet, that they do not use Internet Chat Rooms, and that you use Internet filtering software to block unsuitable material. For more information see the Net Help area of the Usborne Quicklinks Website.

If you don't have use of the Internet, don't worry. This book is a complete, superb, self-contained reference book on its own.

Safety advice

This book contains several activities for you to do at home with your own microscope. Some of the activities involve substances or actions which can be harmful, so you must follow the instructions carefully. Next to certain instructions you will see a caution symbol. This means that you must be very careful and ask an adult to help or supervise you.

⚠️
Caution

THE MICRO WORLD

LARGER THAN LIFE

What you can see with a microscope almost always looks very different from what you can see with your eyes alone. Here are some strange and wonderful things which we would never have known about but for the invention of the microscope.

Plant pollen, shown at 1,100 times its real size.

These tiny water plants live in seas and lakes.

New worlds

Seen with the naked eye, the grain of knotweed plant pollen above looks like a speck of dust.

Under a powerful microscope, you can see that it is a sphere with regular patterns of bumps and ridges on its intricate surface.

Small but vital

A drop of sea water may look as though it has nothing in it, but it could be teeming with life so small you cannot even see it.

The unearthly shapes on the right, for example, are microscopic water plants called diatoms. Despite their tiny size, these plants are vitally important. They are a major source of food for many of the animals that live in the water alongside them.

This ant is pictured here at 125 times its real size.

Alien creature

The creature on the left is called a velvety tree ant. This picture clearly shows the minute hairs that cover its body, making it look furry.

By showing insects close up, microscopes help scientists to explore the alien world of insects. They have shed new light on how the shapes of insects' bodies help them to survive, and the different ways in which they behave.

Medical science

Some of the most important medical discoveries of all time have been made with the help of microscopes. A century ago they were used to prove that most illnesses are caused by germs, which are invisible to the naked eye. The diphtheria germ in the picture below, for example, is shown 54,000 times bigger than it really is.

Household dust viewed with a very powerful microscope.

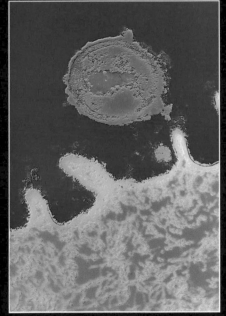

A diphtheria germ on the surface of a person's throat viewed using an extremely powerful microscope.

Fighting disease

As scientists have found out more about germs, they have developed better ways of fighting diseases. For example, diphtheria germs were once a common cause of death. By understanding how the germs behave, this deadly disease can now be treated and even prevented.

Go to **www.usborne-quicklinks.com** for a link to a website where you can take a look at amazing close-up images of insects and see detailed anatomies of an ant, a fruit fly and a mosquito.

Dusty details

Even everyday objects look extraordinary when viewed through a microscope. The picture above, for example, shows some household dust magnified over 28,000 times.

Bits and pieces

This sample contains the tangled remains of a spider's web, scales from a moth's wings, grains of brick dust and insect droppings.

Instead of the grey fluff you can see with the naked eye, the microscope shows that dust is really a collection of minute odds and ends from around the house.

These atoms are among the smallest things that can be seen with a microscope.

Tiny particles

As microscopes become more sophisticated, they can reveal objects that are almost too small to imagine. One type, called a scanning tunnelling microscope, can "see" atoms – the tiny particles that make up everything in the world.

Many millions

The red and yellow mound below is a heap of gold atoms and the green blobs are atoms of a material called carbon. They are shown here many millions of times their real size.

MAKING THINGS LOOK BIGGER

There are two main types of microscopes – optical and electron. Optical microscopes are the kind usually found in homes or schools. Electron microscopes are very complex and expensive machines and are mainly used in medicine and industry.

This picture shows the eye of a needle and a piece of thread, magnified 30 times.

The naked eye

The photograph at the bottom of the page shows a drop of blood as seen by the naked eye, that is, the eye on its own. It looks red because it contains thousands of tiny red blood cells, floating in a clear liquid.

You see the blood when light bounces off it and enters your eyes. You cannot see the individual blood cells because your eyes can only see objects separately that are at least one quarter of a millimetre apart. Any closer together, and the objects seem to blur into one another.

A sample of blood being placed onto a microscope slide.

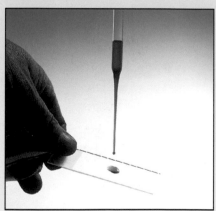

Using light

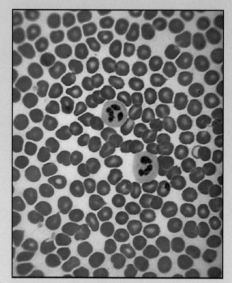

This drop of blood is shown here over 500 times its real size.

The picture above shows a drop of blood viewed with an optical microscope. At this magnification you can see lots of separate red blood cells. (There are also two white cells in the middle of the picture. Find out more about blood cells on page 27.)

Optical microscopes use light and pieces of curved glass, called lenses, to make the object you are looking at appear larger.

Making images

Inside a microscope, an object that you see is actually magnified twice. First, it is magnified by a lens called the objective lens. Then it is magnified again by a second lens, the eyepiece.

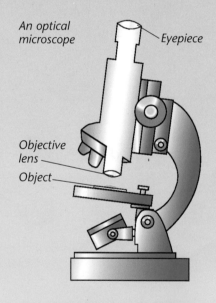

An optical microscope — Eyepiece
Objective lens
Object

How much bigger?

Optical microscopes can magnify between 40 to 2,000 times. They also let you see objects separately that are up to 1,000 times closer together than those that can be seen by the naked eye.

Go to www.usborne-quicklinks.com for links to the following websites:
Website 1 View specimens under an electron microscope and see if you can identify them.
Website 2 See images of a coin magnified by various amounts.

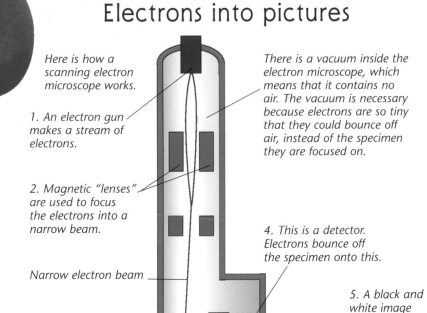

Electrons into pictures

Here is how a scanning electron microscope works.

1. An electron gun makes a stream of electrons.

2. Magnetic "lenses" are used to focus the electrons into a narrow beam.

Narrow electron beam

3. A specimen is placed here. The beam of electrons is moved back and forth over it.

There is a vacuum inside the electron microscope, which means that it contains no air. The vacuum is necessary because electrons are so tiny that they could bounce off air, instead of the specimen they are focused on.

4. This is a detector. Electrons bounce off the specimen onto this.

5. A black and white image forms on a computer screen.

These red blood cells have been magnified 6,000 times by a scanning electron microscope.

Using electrons

Above are some red blood cells which were viewed using a scanning electron microscope*. You can see the actual shape of the cells, as well as details on their surface.

Tiny particles

Electron microscopes use tiny particles called electrons, instead of light, to produce detailed images. These magnified images appear on a computer monitor.

Electron microscopes can see objects that are about one four-millionth of a millimetre apart, and can magnify them by almost a million. This makes them vital tools in areas of scientific research such as medicine. Here, some of the items being looked at, for example certain types of viruses, are extremely small.

Thin slices

Below you can see some red blood cells viewed by a type of electron microscope called a transmission electron microscope. This microscope fires electrons through an extremely thin slice of a specimen, called a section, to show a greatly magnified and detailed view.

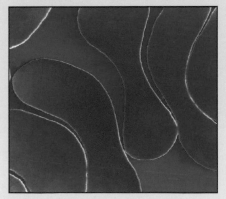

Red blood cells viewed with a transmission electron microscope, magnified 7,200 times.

Computer colour

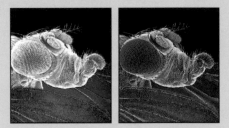

Scanning electron microscope view of a fruit fly. The image on the right has been coloured by computer.

All electron microscope images are black and white, but scientists can add colour to them using a computer. This makes it easier to see the details in the picture.

These pictures are called false-colour images. Most of the electron microscope photographs that appear in this book have been coloured in this way.

Go to **www.usborne-quicklinks.com** for links to the following websites:
Website 1 Use a virtual microscope to zoom in on a flower bud, computer chip and more.
Website 2 View an online slide show that explains how a scanning electron microscope works.

You can see a picture of an electron microscope on page 25.

USING A MICROSCOPE

Most optical microscopes can magnify objects from about 50 up to 1,000 times their real size. The most powerful ones can magnify objects up to 2,000 times. You will find that optical microscopes are easy to use, especially once you know a little about them.

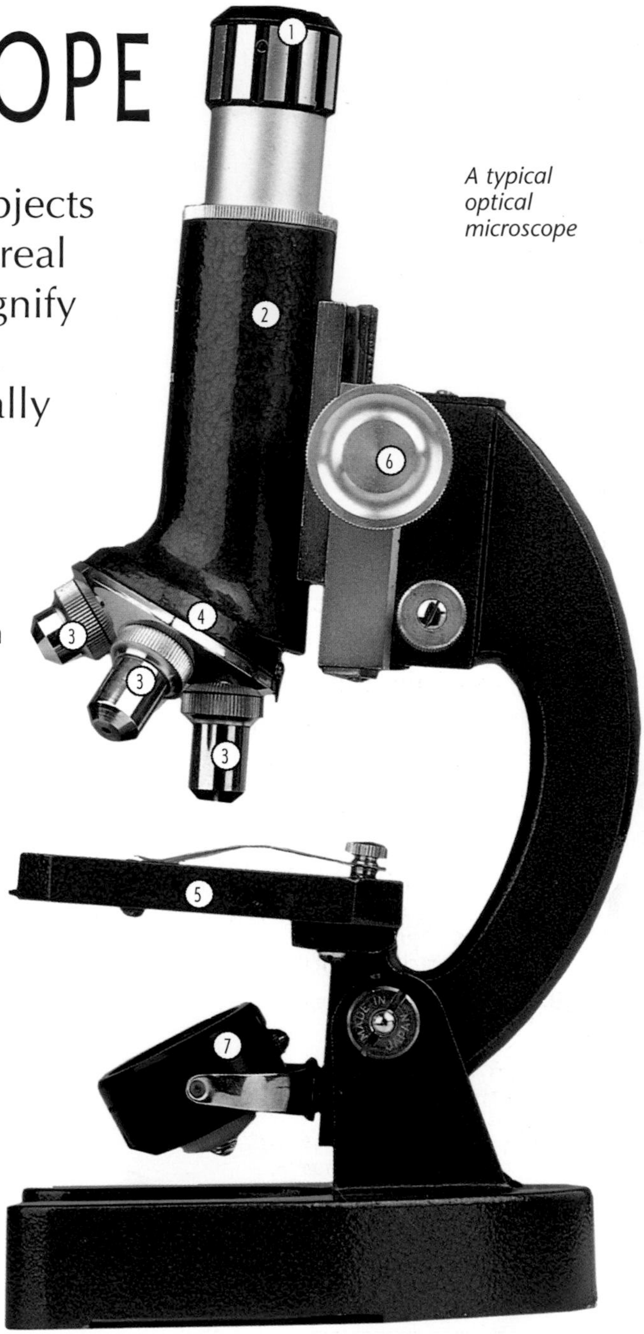

A typical optical microscope

Naming the parts

The picture shows the main parts of an optical microscope.

① Eyepiece. You look through this part. It contains a lens.

② Body tube. On some microscopes this can be tilted.

③ Objective lenses. Most microscopes have three (see *Magnification*, right).

④ Nosepiece. The objective lenses are attached to this.

⑤ Stage. Place the object you want to look at on here.

⑥ Focusing knob. Turn this to make the image sharp and clear.

⑦ Mirror. This reflects daylight or lamplight through a hole in the stage onto the object.

WARNING:
Never point the microscope's mirror at the sun. Its strong rays can reflect off the mirror into your eye and make you blind.

Setting up

Place the microscope on a table so that you can look down the eyepiece easily, without stretching or crouching over. You may be able to tilt your microscope's body tube to make it more comfortable for you to look through the eyepiece.

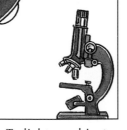

1. To light an object from above (called top lighting), place a lamp 20cm away from the microscope.

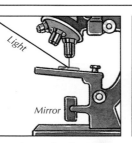

2. Adjust the lamp so its light shines on the object. Alter the mirror so that no light shines through the stage.

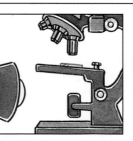

3. To light an object from below (bottom lighting), adjust the lamp so the light shines under the stage.

4. Alter the mirror so that it reflects as much light as possible up through the stage and onto the object.

Feathers are easy to find and interesting to look at through a microscope (see right).

Low to high power

The three pictures on the right show a wing feather viewed with low, medium and high-power lenses. In the first picture you can see the stem-like shaft that runs through the middle of the feather. The comb-like spikes growing out of the shaft are called barbs.

The next picture shows two rows of threads growing from each barb. These threads, or barbules, link together to make the feather strong and flexible, so that the bird can fly. You will see fewer barbs on body feathers, which are for warmth rather than flight.

The third picture shows how barbules attach to each other. The barbules shown in red have tiny hooks. These link into grooves on the barbules shown in yellow.

Shaft of feather

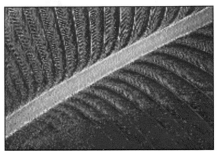

Optical microscope view of a wing feather, seen through a low-power lens.

A closer view of the feather, seen through a medium-power lens.

The feather, seen with a high-power lens.

Try viewing different types of feathers with your microscope.

Magnification

Most optical microscopes have three objective lenses, which magnify the object by different amounts. They are called low, medium and high-power lenses.

A microscope's magnifying power is the power of the eyepiece and the objective lens multiplied together.

This eyepiece magnifies x10.

This objective lens magnifies x40.

Put together, the total magnification is x400.

Focusing

These four steps show you how to focus a microscope. Before you begin, turn the focusing knob to raise the lens as much as possible, then turn the nosepiece to select the lowest-power lens. Always start with this lens as it lets you see more of the object.

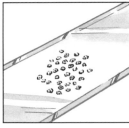

1. Place a specimen (such as the sugar grains shown here) in the middle of a microscope slide.

2. Put the slide on the stage, so the part you want to look at is over the hole. Light the specimen from below.

3. Turn the focusing knob to move the lens close to the slide. Make sure that it does not touch the slide.

4. Look through the eyepiece. Raise the lens until the specimen looks sharp and clear. It is now in focus.

Go to **www.usborne-quicklinks.com** for links to the following websites:
Website 1 Explore an interactive diagram to see how an optical (light) microscope works.
Website 2 Look through a virtual microscope to see objects at different magnifications.

PAPER AND PRINT

We use paper for many purposes, from making books to mopping up spills. If you look at paper through a microscope, you can see that it is made of millions of tiny strands, called fibres. A microscopic view can show why a particular type of paper is suitable for its job.

Piece of newspaper magnified over 40 times. If you look closely, you can see the letters "n" and "d" printed on it.

Flattened fibres

Most paper is made out of fibres of wood from trees such as pine and fir. The wood is cut and mixed with water or chemicals to form a pulp. This is dried on a mesh and pressed to make paper.

A piece of paper, magnified 110 times.

A good way to see paper fibres is to tear the paper and look at the edge under your microscope. In the picture above, you can see how some of the larger fibres were flattened during the paper-making process.

Thick and thin

You can use your microscope to look at types of paper that were made in different ways.

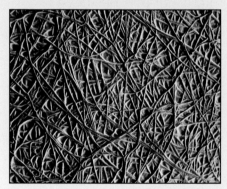

Optical microscope image of writing paper, shown 30 times its real size.

The writing paper above was made from bark-free pulp that was mixed with chemicals. Its long fibres are a similar thickness to each other. They have locked together tightly to form a smooth surface.

A waxy liquid called size has been added to the paper. This makes the surface less absorbent, so ink does not soak into the holes and smudge the writing.

Short strands

Newspaper pulp is made from tree trunks that have been ground up with their bark still on, then mixed with hot water. If you look at a torn edge of newspaper under your microscope, you will see that the fibres are different lengths and thicknesses, stuck together in a random way.

Printing inks

Newspaper does not contain size, which means that it absorbs water easily. But the printer's ink does not smudge because it is made of alcohol, which dries very quickly, before it soaks into the paper.

① ② ③

Go to www.usborne-quicklinks.com for links to the following websites:
Website 1 Learn more about paper and look at SEM images of paper fibres.
Website 2 Follow step-by-step instructions on how to make paper.

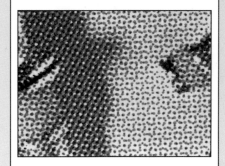

Practical papers

The large, green bundles below are fibres of soft toilet tissue paper. If you look at a piece of toilet tissue under a low-power microscope, you will see that the fibres are widely spaced. Liquid flows easily into the spaces, making tissue paper very absorbent.

Piece of re-usable sticky note paper.

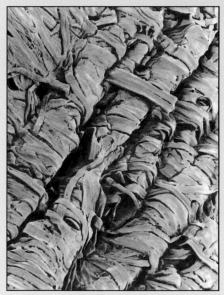

Sample of toilet tissue, magnified 110 times by an electron microscope.

You could try looking at different types of paper with your microscope.

1. Kitchen paper towel
2. Crepe paper
3. Writing paper
4. Silk paper
5. Brown envelope paper

Stuck up

Above you can see the surface of a re-usable sticky note, viewed with an electron microscope. It has been magnified 450 times.

Inside each yellow bubble is a small amount of glue. The surface of the bubble is made from a material that breaks easily under pressure.

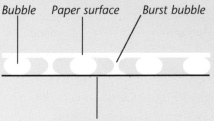

Bubble   Paper surface   Burst bubble

Hard surface

Each time you press a piece of the notepaper to a surface, some bubbles burst, releasing tiny amounts of glue. These stickers can be used several times before all the glue bubbles are used up.

Going dotty

This magazine picture has been magnified eight times. You can see that it is made up of tiny coloured dots.

Only four colours are used, but when you look at the picture from farther away, it seems to contain many different colours.

All the pictures in this book are made up of tiny dots in these four colours.

Cyan   Magenta   Yellow   Black

FIBRES AND FABRICS

Many of the clothes we wear are made from fabrics such as cotton or nylon. These fabrics are woven from threads called yarns, which are made of microscopic fibres. On these two pages you can see why certain types of fabric feel and look different.

Electron microscope view of fibres from a sleeping bag, magnified 750 times.

Rough and smooth

In the electron microscope picture below you can see two different types of fibres.

The rough, green fibre is cotton, which comes from the seeds of a plant. The smooth, yellow strands are fibres of a man-made material called polyester. Man-made fibres are often smoother than natural ones.

These cotton and polyester fibres have been magnified 1,500 times.

Woven yarns

The picture below shows how cotton yarns have been loosely woven together to make a fabric.

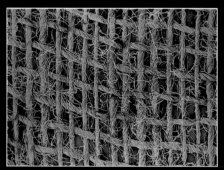

This piece of fabric has been magnified 15 times by an electron microscope.

The fabric above is called cheesecloth. If you look closely at the picture, you can see that each thread in the fabric is itself made up of individual cotton fibres.

Warm wool

Below you can see some fibres of undyed sheep's wool. These thick fibres contain hollow pockets of air. When you wear clothes made from wool, your body heat warms the trapped air, which then helps to keep you warm.

Optical microscope image showing wool fibres magnified 960 times.

The surface of each fibre is covered with tiny scales which point toward its tip. When you wash wool cloth, these scales rub against each other. The combination of heat, water and rubbing can cause the fibres to curl up, making the cloth shrink.

Go to **www.usborne-quicklinks.com** for links to the following websites:
Website 1 Explore different fibres and fabrics and see denim, fleece and silk close-up.
Website 2 Compare natural, synthetic and ceramic fibres under the microscope.

No sweat

Holes in fabric can help to keep you dry, as well as warm. The picture below shows a piece of fabric from a rainproof coat. Its surface has been coated with a layer of man-made rubber which stops rain from soaking through.

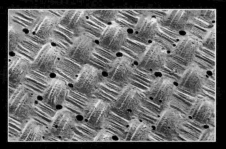

Waterproof fabric used in outdoor clothing, magnified 30 times.

The tiny holes that you can see are too small for drops of water to soak through. But they do allow damp air and water vapour from your body to escape through them.

Fabric

Raindrops cannot soak through.   *Water vapour escapes.*

This prevents you from getting too sweaty and uncomfortable inside your coat.

Soft to the touch

The man-made fibres at the bottom of the page are used to make a fabric called microfibre. This feels very soft, like the skin of a peach, and is often made into clothes. You can see a magnified view of microfibre cloth immediately below.

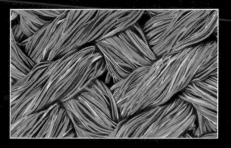

A piece of microfibre fabric shown at 60 times its real size.

Microfibre is ideal for cleaning glasses or other lenses. This is because its wedge-shaped fibres scoop up the layer of oil and dirt that builds up on the lenses. The dirt is trapped between the strands, so the cloth does not smear the lens.

Ends of microfibre fibres, magnified 5,000 times.

Pockets of air

These colourful shapes are fibres of a man-made material called Dacron* polyester. It is used as padding in sleeping bags. There are holes through the middle of each fibre which trap air, in the same way as the air pockets in wool fibres do. This helps to make the sleeping bag snug and light.

Air fills these holes.

Microscopic holes

To make these fibres, liquid polyester is forced through a metal plate, with hundreds of intricately shaped holes in it.

Fibres with large holes hold more air but this air is easily squashed out when you lie on them. Smaller holes hold less air, but the fibres keep their shape better. Sleeping bags contain a mixture of fibres, to make them as warm as possible.

*Dacron is a registered trademark of DuPont.

ODDS AND ENDS

There are plenty of odds and ends you can find around the house, which are fascinating to look at under a microscope. Here are some close-up views of everyday objects that show you what they are made of and reveal why they work the way they do.

Nylon netting

The picture below shows an electron microscope view of a piece of fishing net, for use with indoor fish tanks. It has been magnified 88 times.

The yellow yarn is nylon, a man-made material, which has been woven into a fine mesh. You can see how smooth the nylon fibres are compared with natural fibres such as cotton and wool (see page 14).

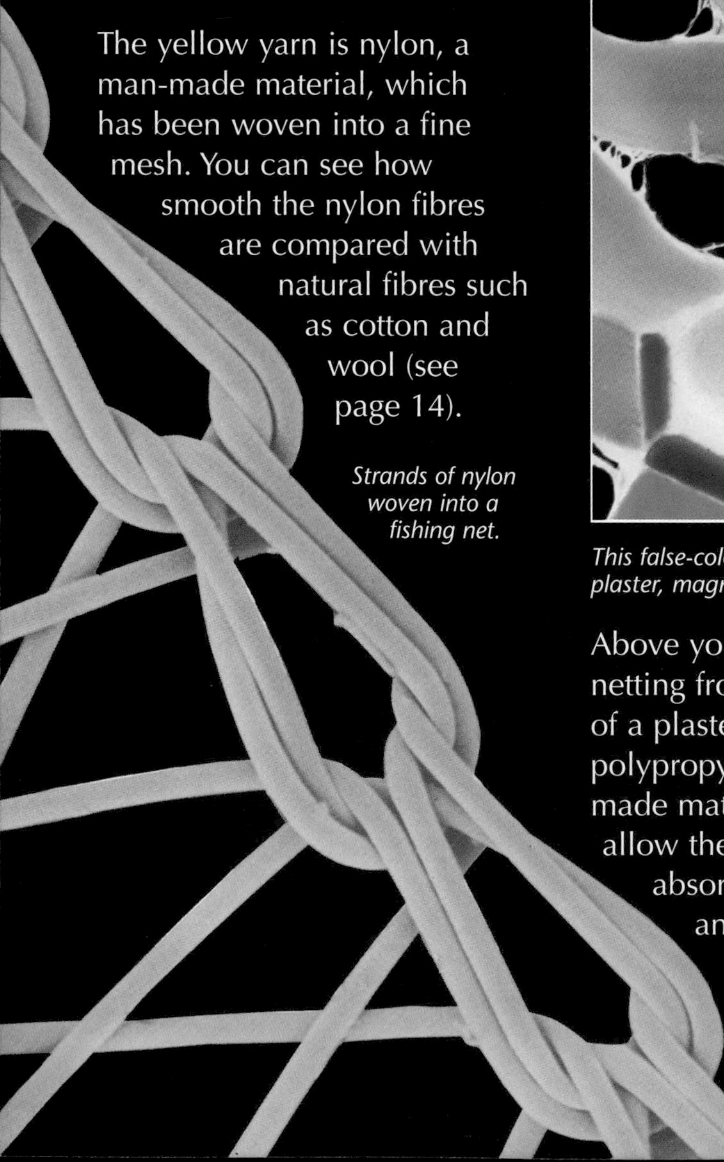

Strands of nylon woven into a fishing net.

Non-stick netting

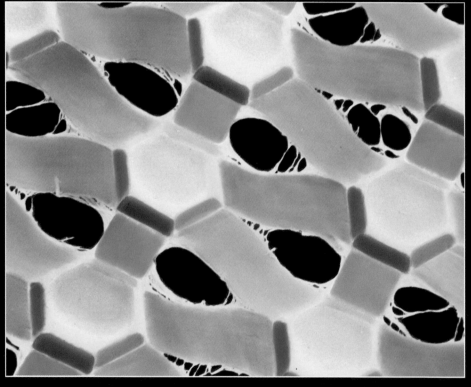

This false-coloured electron microscope image shows a piece of netting in a sticking plaster, magnified over 40 times.

Above you can see a piece of netting from the dressing pad of a plaster. It is made out of polypropylene, another man-made material. The holes allow the dressing pad to absorb moisture, such as antiseptic cream or blood from a wound.

The polypropylene netting does not absorb moisture. This means that it helps to stop the dressing pad from sticking to the wound.

Dressing pad
Polypropylene net
Wound

Man-made fibres of different widths have been woven together to make this pan scourer.

Soft scourers

The electron microscope image above shows the surface of a scouring pad that can be used for cleaning pans. It has been magnified over 70 times.

Gentle cleaning

Large, yellow ribbons of nylon have been woven into the fabric. These hard-wearing strands scrub off food that has burned onto pans. The fibres' rounded shape means that they clean the pan without scratching its surface.

Firm fasteners

Below you can see an electron microscope view of a piece of hook and loop fastener, which you might use to fasten clothes, sports shoes or bags. It is made out of nylon, and is manufactured in two separate parts.

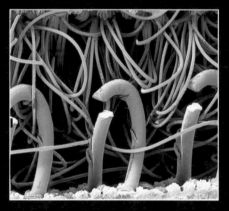

A piece of hook and loop fastener, shown here magnified eight times.

Loose strands

The upper part of the fastener (pictured in green) is made up of a series of loops. These loops are loose strands in the otherwise tightly-woven nylon fabric.

The hooks, shown in blue, are thicker loops of nylon that have been woven into the fabric, then cut to form hooks.

When you bring the two pieces together, the loops catch on the hooks. They form a bond which is strong but is easy to peel apart.

Micro drawings

A good way to keep a record of what you see through your microscope is to draw it. Before you start, draw a saucer-sized circle on a piece of paper. Then lightly draw a grid over the circle, making sure that the lines are evenly spaced.

1. Position the paper on the table so that you can see it without raising your head from the eyepiece.

2. Imagine a grid over the object. Notice where on the imaginary grid each part of the object is.

3. Draw it in the matching area of the paper grid. Keep your drawing simple.

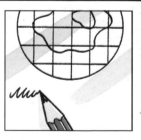

4. Remember to write a title and the date on your drawing, and the total magnification you have used.

Go to **www.usborne-quicklinks.com** for links to the following websites:
Website 1 Take an online quiz and see if you can identify everyday items by examining lots of close-up images.
Website 2 See close-ups of odds and ends, such as animal hair, sand, salt and sugar crystals and sewing needles.

CLUES TO THE PAST

Much of what we know about our ancestors comes from looking at the remains of things they left behind. Scientists, called archaeologists, examine this evidence of past ways of life. Some remains are extremely small, so they study them with microscopes.

These wild seeds are from plants that grew hundreds of years ago.

Pot pieces

Man-made products such as bricks, tiles and pottery can provide archaeologists with lots of clues, because they do not rot away over time. The picture below, for example, shows a fragment of building material which was used to construct a Roman bath house in 450BC.

Archaeologists use microscopes to look at pottery pieces. A specialist can find out what a pot was used for. For instance, a high sand content in pottery helps it to withstand heating up without cracking. So a fragment containing a lot of sand grains may have come from a cooking pot.

Optical microscope image of Roman building material, magnified over 30 times.

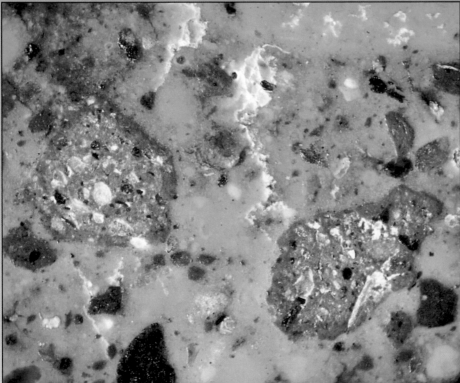

Food for thought

The minute remains of plants can be preserved in the ground. If they are found in places where people once lived, they might offer clues about what those people ate. The raspberry and blackberry seeds above, for example, were discovered in a city toilet pit dating from medieval times (the eleventh to the fifteenth century).

The charred oat grains below were found next to a buried medieval fireplace in London. They may have fallen there when they were being cooked. Finds like these show archaeologists the kind of food that was brought into London to feed its growing population.

Microscopic examination of these grains has shown that they are oats.

Go to **www.usborne-quicklinks.com** for links to the following websites:
Website 1 Find out how the British Museum uses microscopy in its archaeological investigations.
Website 2 See how archaeologists use microscopes and other tools to discover clues about the past.

Preserved pollen

The specks of grain pollen* in the picture below are about 5,000 years old. They were found in Britain and show that people had started farming there by this time.

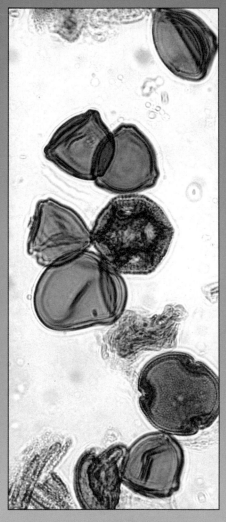

This 5,000 year-old pollen fell on marshy ground. This preserved it so well it is still possible to identify individual types.

Although the sample was collected from flat, marshy ground, it contains pollen from trees as well as pollen from crops that were grown for food. This suggests that farmers chopped down forests to clear the land for growing their own crops.

*You can find out more about plant pollen on pages 44-45.

Ancient animals

Microscopic animal remains can give archaeologists clues about how the land and climate have changed through the ages. They also reveal more about how people lived long ago.

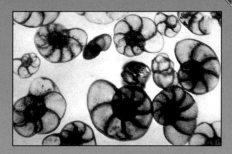

These are the shells of creatures that lived in sea water.

The delicate shells above are from tiny water creatures called foraminifera. They live in sea water. If archaeologists find creatures like this in soil, it suggests that the area was once below the sea.

Freshwater fleas

The creature on the right is a cladoceran, sometimes known as a freshwater flea. If creatures like this are found in a soil sample, it can indicate that the land was once covered in fresh water, such as a lake, ditch or moat.

This freshwater creature is shown 75 times larger than its real size.

This electron microscope image shows the head of a dog flea, which was found inside a Roman home.

Pets and pests

Above you can see the head of a dog flea. It was found with other remains from a house floor in a Roman settlement at York, in the north of England.

Cats and dogs

People kept dogs to work with them, doing jobs such as hunting, and rounding up sheep. The dog flea, found in dirt swept from the floor, suggests that the dogs lived in the house with their owners.

Surprisingly, archaeologists have not found any evidence of ancient cat fleas in this Roman settlement. They think that this may be because people kept their cats outside, where they had to kill rats and mice for food.

SOLVING CRIMES

Even the most careful criminal can leave behind minute clues. These fragments of evidence are often too small to be seen with the naked eye, so microscopes play a vital part in helping the police to solve crimes.

Looking for clues

Police scientists use optical microscopes to examine all sorts of possible evidence for clues. This helps them to decide what to do next. For example, they could examine the evidence further with an electron microscope or perhaps take samples, such as blood or soil, for other tests.

Matching fibres

When a crime such as a burglary or murder takes place, particles of the culprit's clothing, such as wool fibres from a sweater, are often left behind. If the fibres found at the crime scene match ones on clothing which belongs to a suspect, they can help police to link the person with the crime.

Other clues

Other things that police scientists might examine for clues include hairs, carpet fibres and fragments of glass from broken windows.

Microscopes on wheels

To inspect large objects, such as cars, police scientists use an operating microscope*. This is a microscope attached to a hinged arm on a tall stand. The stand is wheeled up to the object and the microscope is tilted so that scientists can examine the object from different angles. They can then decide which parts to look at more closely.

Electron microscope image of layers of paint on a car, magnified 40 times.

Car thieves sometimes respray stolen cars to disguise them. But careful examination with a microscope of a suspected stolen car can help police to identify it. In the picture above, for example, you can see a section of bodywork from a rusty car. Although the car is blue, this picture shows that it was originally green.

Watery clues

Microscopes are also used to look for clues in bodies that have been found in suspicious circumstances. For example, if microscopic water plants are found inside the lungs of a body taken from a river or lake, it is likely that the person drowned. If no water plants are present, the victim was probably dead before entering the water.

Guns and guilt

Microscopes can help to solve crimes involving guns. The inside of all gun barrels have tiny grooves cut into them. These make the bullet spin, which helps it to travel straight and hit its target.

When a gun is fired, the bullet scrapes along the inside of the barrel. The grooves scratch the side of the bullet and make a mark which is as individual as a human fingerprint. These microscopic marks can be compared with the marks on other bullets to see if they were fired from the same gun.

The scratches on these bullets do not match, so they were fired by different guns.

Go to **www.usborne-quicklinks.com** for links to the following websites:
Website 1 Find out more about methods used by criminal and historical investigations.
Website 2 Take a look at some of the forensic services of the FBI .

*You can see a picture of an operating microscope on page 38.

THE HUMAN BODY

ON YOUR HEAD

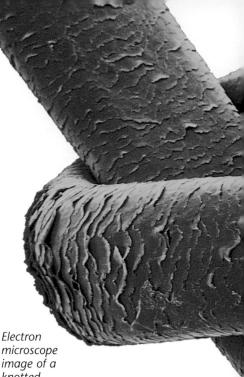

What could be more familiar than your own face? You probably look at your reflection every day, when you comb your hair or clean your teeth. Here are some less familiar views of your hair, teeth and tongue, all shown massively magnified by an optical or electron microscope.

Overlapping scales

The brown, snake-like twist at the top of the page is really a knotted strand of human hair. It has been magnified over 750 times by an electron microscope. The outer surface of the hair is covered with overlapping scales.

Hair is made from a substance called keratin, which is also found in fingernails and toenails. In the picture below you can see a hair that has been cut down the middle. It is made up of three different layers.

Optical microscope image of a slice of human hair, magnified 490 times.

The outer layer, or cuticle, is made up of thin, flat scales of keratin. This layer is harder than the others, and it helps to protect the hair.

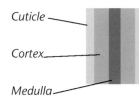

Cuticle

Cortex

Medulla

Each hair has three layers.

The next two layers, the cortex and the medulla, give the hair its colour. This is because they both contain tiny grains of a chemical called melanin.

Black to white

Black hair contains mostly pure melanin. Blond and red hair contain melanin mixed with other chemicals, such as iron and sulphur. As you grow older, these grains are gradually replaced with pockets of air, making your hair turn white.

Electron microscope image of a knotted human hair.

Tiny holes

The pink, fingertip-like shapes below are young hairs that have started to push their way out of the scalp.

Electron microscope view of hairs growing from the scalp. Shown 125 times their real size.

Each hair grows from a tiny hole called a follicle. There are thousands of follicles in your scalp, and many more on the rest of your body. Only the skin on the palms of your hands and the soles of your feet has no follicles or hairs.

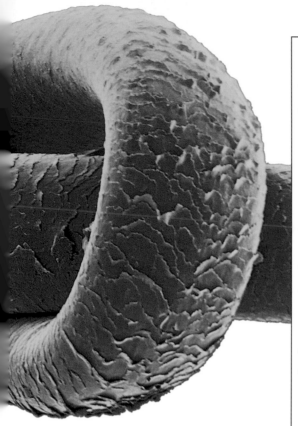

Tongue and taste

The surface of a human tongue is covered with tiny lumps called papillae. There are various types of papillae, each with a different job to do. For example, the papilla you can see at the bottom of the page contains taste buds.

Taste buds are sensitive to four main tastes: sweet, sour, salty and bitter. These tastes can be detected all over the tongue, but some areas of the tongue are thought to be more sensitive to certain tastes.

Map of the tongue showing where the different tastes are sensed more strongly.

Bitter

Sour — — Sour

Salty — — Salty

Sweet — — Sweet and salty

Electron microscope view of the surface of the tongue, magnified 420 times.

The pointed shapes above are another type of papilla. These form a rough surface which helps the tongue to grip food and move it around your mouth while you are chewing.

The tip of each papilla is made out of keratin. This helps to make them tough and hard-wearing.

Hard surface

Below you can see the highly magnified surface of a tooth. It is covered in enamel, a hard substance which enables the tooth to bite without breaking, and protects it from decay. The regular rows are layers of calcium, which is the main ingredient of enamel.

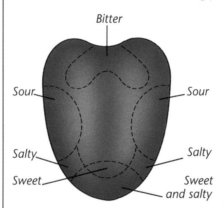

This papilla on the tongue contains taste buds under its surface. It is shown magnified nearly 500 times.

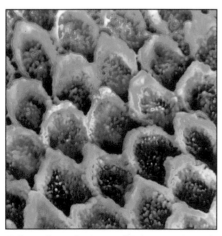

False-colour electron microscope view of the hard enamel covering of a tooth.

Go to **www.usborne-quicklinks.com** for a link to a website where you can take a visual journey inside a single strand of hair and discover lots of fascinating facts including why it's so strong.

BODYWORK

The human body is a complex collection of different parts, each with its own job to do. Here you can see what some body parts look like through a microscope, and find out more about how they work.

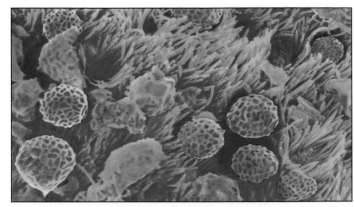

In this picture of a human windpipe, you can see grains of pollen (orange) and dust (brown) which have been breathed in.

Air conditioning

Every time you breathe in, air travels from your mouth or nose, down a tube called the windpipe, and into the lungs. The picture below shows an electron microscope view inside a human windpipe. It has been magnified about 9,000 times.

The blue blobs in the middle of the picture are known as goblet cells. They produce a thick, sticky liquid called mucus, which traps dirt such as pollen and dust, and stops it from reaching the lungs.

Runny noses

The hair-like threads, called cilia, gently waft the mucus away from the lungs, toward your nose and throat. When you have a cold, your body makes more mucus to carry germs away from your lungs.

False-colour electron microscope view of the lining of a human windpipe.

Sturdy bones

Microscopes have revealed that bones are made up of several different layers.

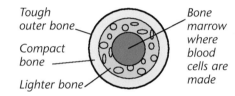

Tough outer bone
Compact bone
Lighter bone
Bone marrow where blood cells are made

The picture at the bottom of the page shows the layer called compact bone. It is a dense material, arranged in rings around channels containing blood vessels. This layer strengthens the bone.

Compact bone forms tubes which run along the length of the bone, making it stronger.

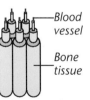

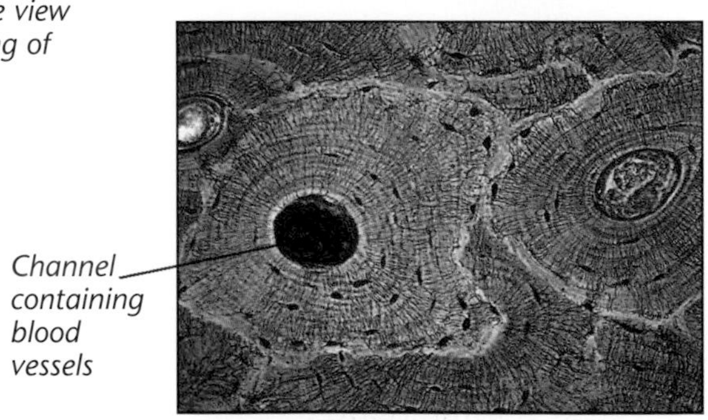

Blood vessel
Bone tissue

Channel containing blood vessels

A magnified section of compact bone.

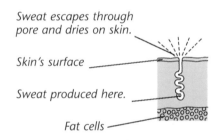

Body pictures

Most of the body parts on these pages were viewed using a scanning electron microscope* like the one shown below. A beam of electrons scans the surface of a sample inside the tall, metal column. A picture of the sample then appears on the video screen.

A scanning electron microscope

Specimens for electron microscopes need to be carefully prepared. A small sample of tissue, such as skin, is removed from the body. It is then frozen so that it does not lose its shape in the vacuum inside the electron microscope.

Gold coating

Normally, electrons would pass straight through a specimen, so its surface is coated with a very thin film of gold. This makes the electrons bounce off the specimen, creating the amazing 3-D images you can see on this page.

Warm and cool

The yellowy-brown bubbles in the picture on the right are from a layer of fat which lies beneath the skin. This layer acts like a sort of blanket, helping to keep your body warm.

Protective patchwork

In the picture below you can see a magnified view of human skin. Its surface is a patchwork of skin cells. These fit together to form a layer which protects your body against germs, heat and cold.

Cooling down

The blue beads are drops of sweat, which contain salt and water from your body. Sweat is produced beneath the skin, and oozes up to the surface through tiny holes, or pores.

This layer of fat helps to keep your body warm. Shown magnified 1,300 times.

As the sweat dries on your skin, it uses up heat from your body, cooling you down.

Sweat escapes through pore and dries on skin.

Skin's surface

Sweat produced here.

Fat cells

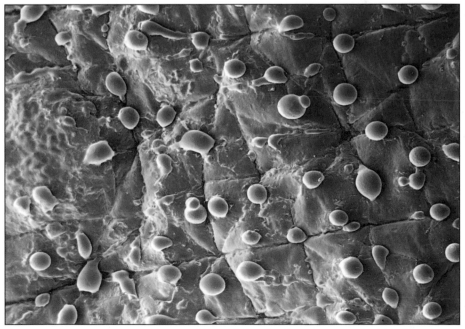

Electron microscope image of human skin with beads of sweat, magnified 30 times.

*Go to **www.usborne-quicklinks.com** for a link to a website where you can look at a fascinating selection of magnified pictures of human tissue and organ systems.*

You can find out more about how a scanning electron microscope works on page 9.

BODY CELLS

Your body is made out of about 100 million million tiny units called cells. These two pages show a microscopic view of some of the different kinds of cells that are found inside you.

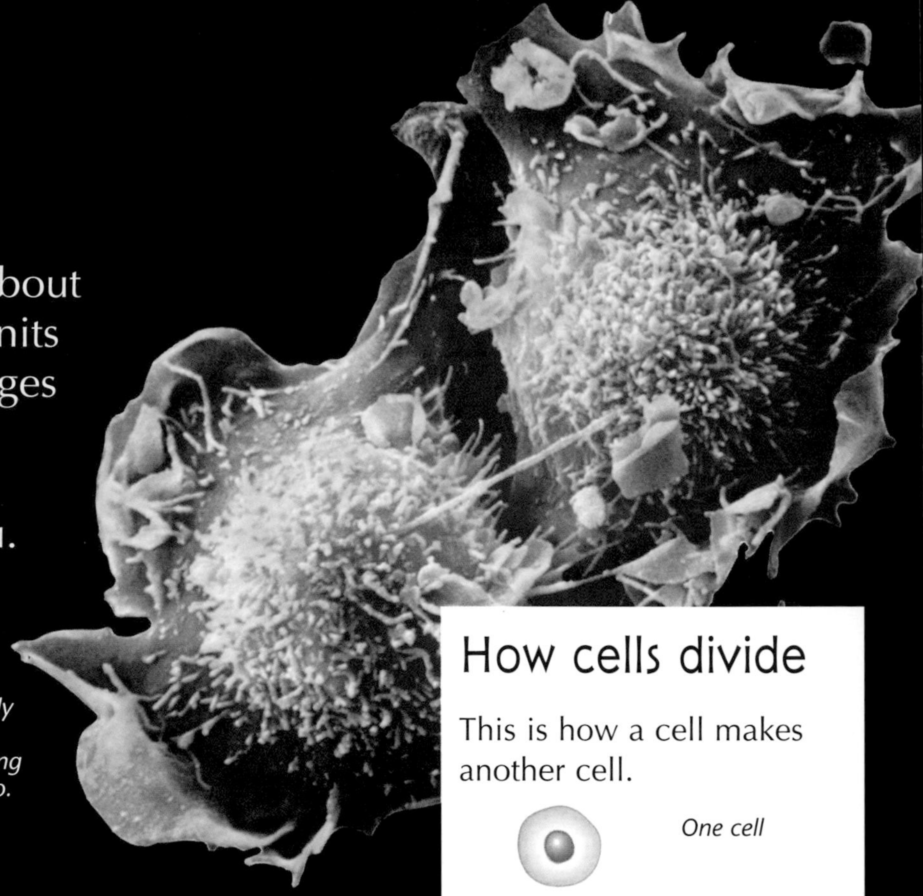

A body cell dividing in two.

Shaped for action

The picture below shows some nerve cells viewed with a microscope. The hair-like endings pick up signals in one part of the body that are then carried by long, thin fibres to other parts of the body.

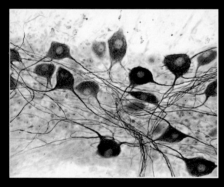

These nerve cells pass messages around your body about sensations such as pain.

New cells for old

The electron microscope picture above shows a cell making a new cell by dividing into two. Body cells make new cells because millions of cells die every second, and need to be replaced. Cells divide in several stages: you can see these stages in the diagram on the right.

How cells divide

This is how a cell makes another cell.

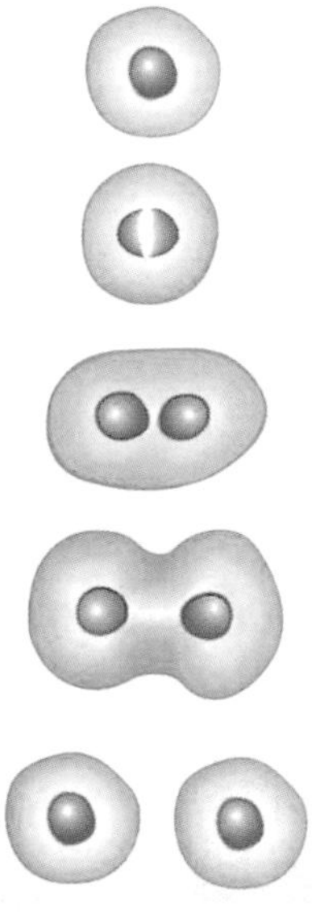

One cell

Cell's nucleus (control centre) starts to divide.

Two identical nuclei form.

Cell narrows in the middle.

Cell splits in two. The new cells are exactly the same as the original one.

Cheek cells

You can look at body cells after taking them from the inside of your cheek. Use your microscope to inspect the cells. They will look wide and flat.

Never share a cotton bud with another person. Keep your own cheek cells to yourself.

1. Collect some cheek cells by gently wiping a clean cotton bud around the inside of your cheek.

2. Use the cotton bud to spread the cells onto a microscope slide. Throw away the cotton bud.

3. Carefully place a cover slip over the cheek cells. Look at the cells, lighting them from below.

4. Soak the slide overnight in disinfectant. Then clean it using warm water with washing-up liquid.

Go to **www.usborne-quicklinks.com** for links to the following websites:
Website 1 *Watch an amazing detailed animation of cell growth and division.*
Website 2 *Explore an image gallery for more pictures and information about body cells.*

Protective cells

The optical microscope picture below shows how cheek cells fit together tightly, like pieces of a jigsaw.

These cells, and others which form a protective covering, such as skin or the lining of the stomach, are called epithelial cells.

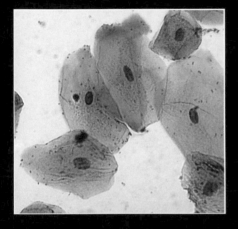

These cheek cells have been dyed to make them easier to see.

Blood cells

To the naked eye, blood looks like a red liquid. With a low-power microscope, scientists can see that it is a mixture of cells floating in a straw-coloured liquid called plasma.

With a powerful microscope, scientists can clearly see the three types of cells found in blood: red cells, white cells and platelets.

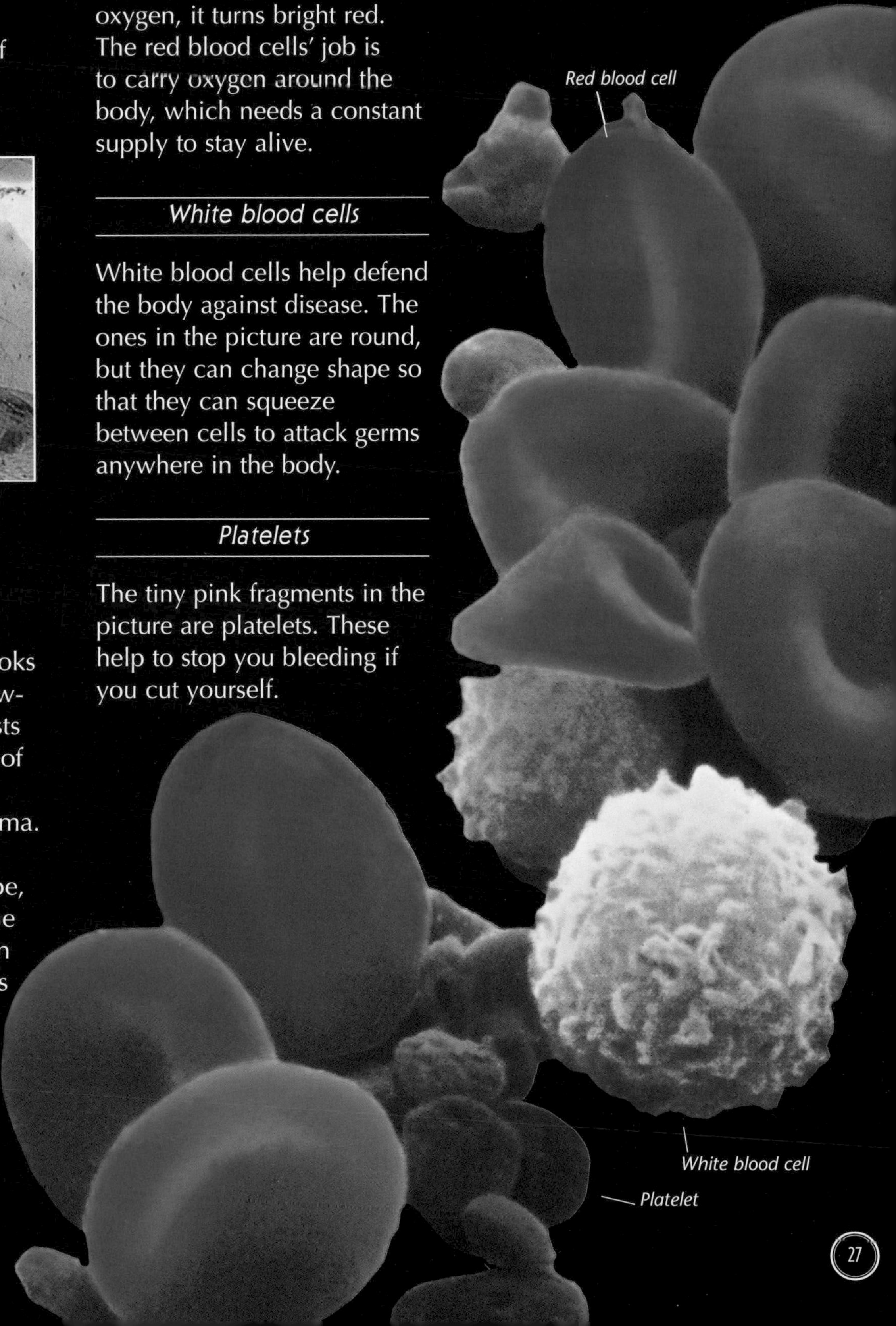

This false-colour electron microscope picture shows three different types of cells in a drop of blood.

Red blood cells

The bright red, doughnut-shaped cells in the picture are red blood cells. They contain a chemical called haemoglobin. When this mixes with a gas called oxygen, it turns bright red. The red blood cells' job is to carry oxygen around the body, which needs a constant supply to stay alive.

White blood cells

White blood cells help defend the body against disease. The ones in the picture are round, but they can change shape so that they can squeeze between cells to attack germs anywhere in the body.

Platelets

The tiny pink fragments in the picture are platelets. These help to stop you bleeding if you cut yourself.

How many?

A drop of blood the size of a pinhead contains about five million red blood cells. It also contains around 7,500 white blood cells and over 250,000 platelets.

Red blood cell

White blood cell

Platelet

INSIDE A BODY CELL

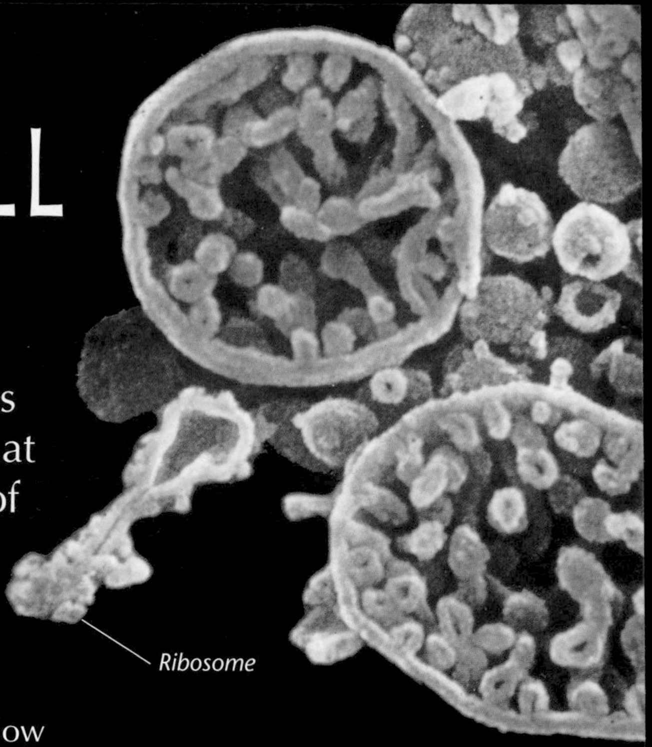

The inside of a body cell is surprisingly complicated, as each cell is made up of several distinct parts. Modern microscopes have helped scientists to discover exactly what each part does, and here you can see some of the parts in detail.

Ribosome

What is in a cell?

The optical microscope image below shows a single cheek cell. In it you can clearly see three main features that can be found in most body cells.

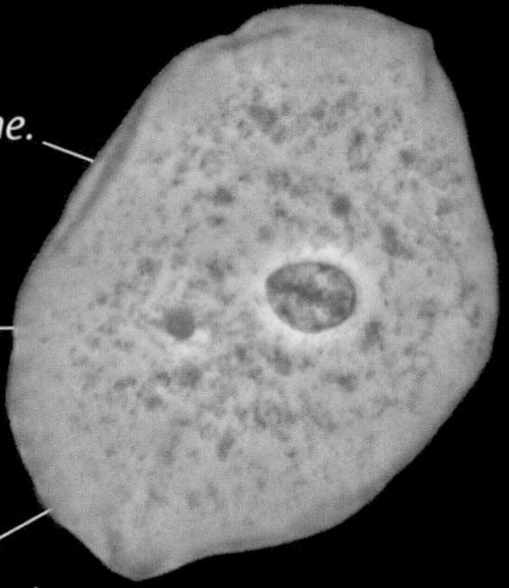

The edge of the cell is called the cell membrane.

The dark patch in the middle is called the nucleus.

The rest of the cell is called cytoplasm.

The cell membrane forms a protective layer around the cell. It also holds the contents of the cell together.

The nucleus controls everything that happens inside the cell.

The watery cytoplasm contains minute parts called organelles. Each organelle has a particular job to do to keep the cell working.

A typical cell

The circular diagram below shows the organelles inside a typical cell. You can identify some of these parts in the electron microscope picture of a liver cell at the bottom of the page.

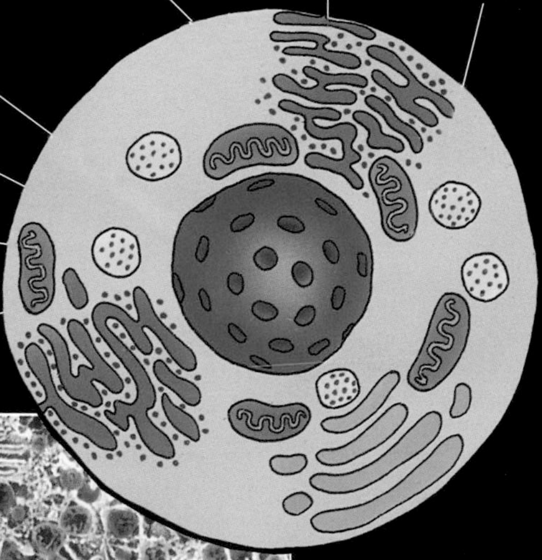

Endoplasmic reticula. Materials move around the cell along these passages.

Ribosome. The cell is made from proteins which are made on ribosomes like this one.

These blue segments make up the Golgi complex. Proteins are stored here.

Nucleus

Cell membrane

Lysosome. This destroys old and diseased parts of the cell.

Mitochondrion. This makes energy for the cell.

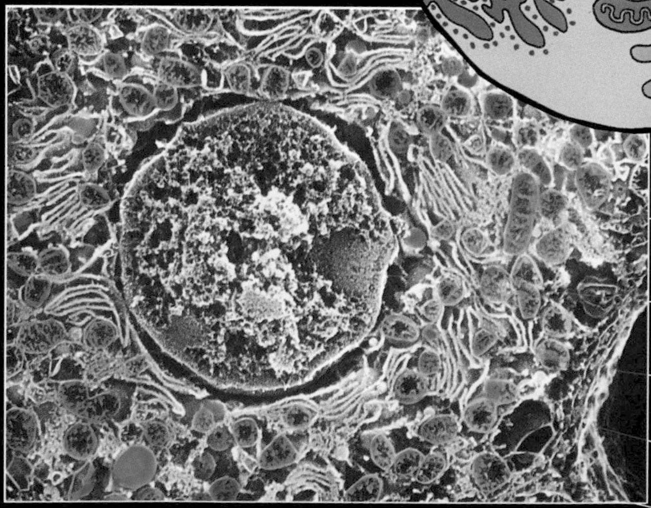

Endoplasmic reticulum

Nucleus

Mitochondrion

Cell membrane

Ribosome

Electron microscope view inside a liver cell.

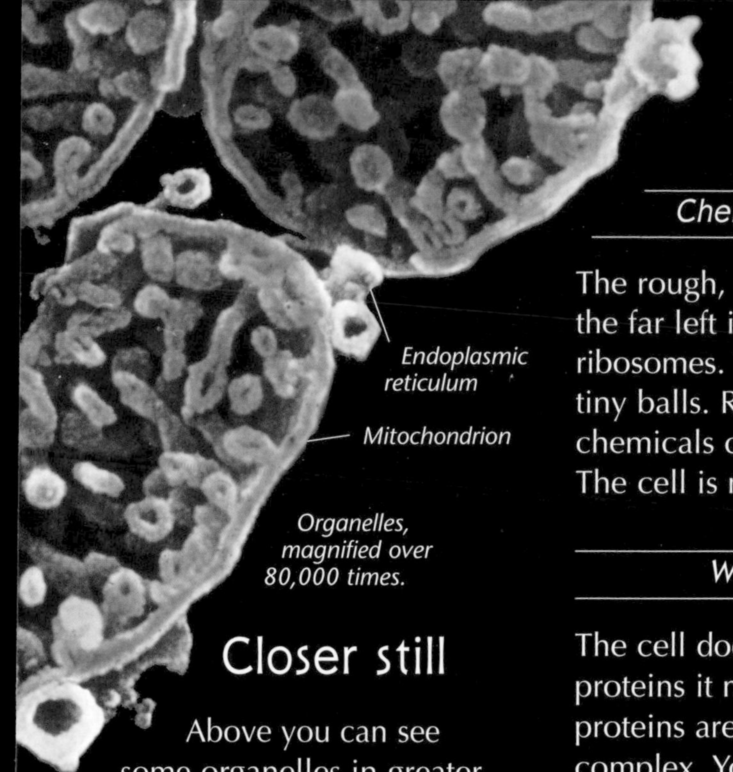

Endoplasmic
reticulum

Mitochondrion

*Organelles,
magnified over
80,000 times.*

Closer still

Above you can see
some organelles in greater
detail. The pink shapes are
mitochondria. Food and
oxygen mix in them to make
energy to keep the cell alive.

The more active a cell is, the
more mitochondria it needs.
The liver cell on the left works
hard changing food into energy.
It has plenty of mitochondria
to enable it to do this.

Transport channels

Alongside the mitochondria
you can see yellow channels.
These are endoplasmic
reticula, which help materials
to move quickly around
the cell.

Chemical factory

The rough, yellow loop on
the far left is studded with
ribosomes. They look like
tiny balls. Ribosomes make
chemicals called proteins.
The cell is made out of these.

Warehouse

The cell does not use all the
proteins it makes. Spare
proteins are stored in its Golgi
complex. You can see one
close up in the picture below.

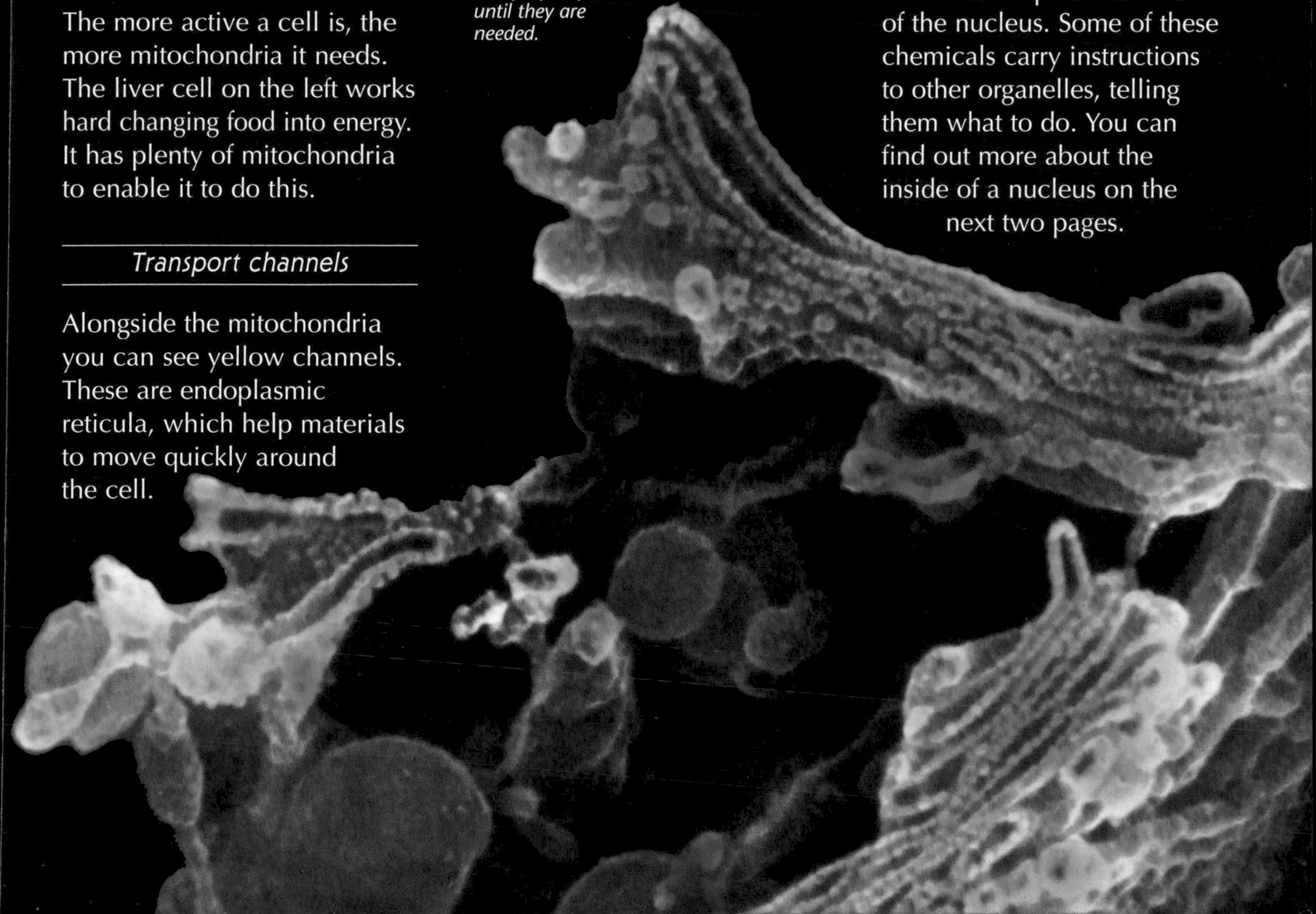

*Electron microscope picture of a Golgi
complex. Proteins are stored in the pink
and purple spheres
until they are
needed.*

Headquarters

The red area in the picture
below is the cell's control
centre, or nucleus. This
nucleus is sphere-shaped,
but some others are oval.

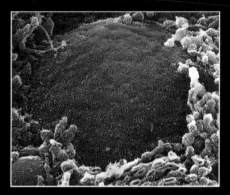

*A false-colour electron microscope
photograph of a nucleus.*

The tiny holes you can see
on the surface allow certain
chemicals to pass into and out
of the nucleus. Some of these
chemicals carry instructions
to other organelles, telling
them what to do. You can
find out more about the
inside of a nucleus on the
next two pages.

INSIDE A NUCLEUS

At the middle of most body cells lies a nucleus. Not only does it tell the cell what to do, it contains a complex set of instructions which make up the recipe for human life. Here is what the microscope reveals as it looks deep into the nucleus.

Nucleus

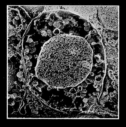

The large, brown oval on the left is a cross-section* of a nucleus inside a body cell, viewed with a scanning electron microscope. The nucleus is the largest part inside the cell.

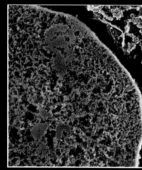

Inside a nucleus, magnified by an electron microscope.

The pink, rounded shape above is a closer view inside a nucleus. Like all the other images on these pages, colour has been added to make it easier to see the detail. The denser pink areas that you can see are called nucleoli. Each nucleolus makes ribosomes like the ones described on pages 28 and 29.

Chromosomes and genes

Hidden among the lumpy fibres in the nucleus are bundles of tightly-packed threads, called chromosomes. They contain thousands of tiny units called genes. These control the cell and hold instructions for a new human life.

Most cells in your body contain an identical set of chromosomes. The picture below shows an electron microscope image of two chromosomes, magnified more than 10,500 times.

These two bundles of threads are individual chromosomes.

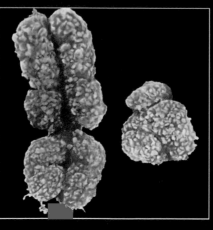

A strand of DNA, shown at seven million times its actual size.

Complex chemical

Chromosomes and genes consist of a very complex chemical called DNA. You can see a length of DNA unwinding from the bottom of the larger chromosome on the left. The picture above shows a massively magnified single strand of DNA.

Go to www.usborne-quicklinks.com for a link to a website where you can zoom in on a human hand and find out what happens inside the nucleus of almost every cell, and see animated images of DNA.

*A cross-section is a slice through something.

DNA strands

DNA is too small to be seen with an ordinary electron microscope. The picture of DNA below was taken using a more powerful instrument called a scanning tunnelling microscope. It shows objects as outlines rather than solid shapes.

Genes

The genes that are found in your chromosomes are very important because they are responsible for making everyone slightly different. For example, your eye and hair colour, and the pattern of your fingerprints, are all controlled by your genes.

This sperm cell and egg cell are joining together to start a human life.

Sharing information

You inherit your genes from your parents. Using powerful microscopes, scientists have been able to discover exactly how this information is passed from parents to their children.

Chromosomes and their genes are passed on during a process called fertilization. This happens when an egg cell from a woman's body joins with a sperm cell from a man's body.

Sperm cell   *Egg cell*   *Fertilized egg cell*

You can see the actual moment when the sperm joins the egg in the photograph at the top of the page.

Making babies

In the electron microscope picture above, the rounded shape is an egg cell and the tadpole shape is a sperm cell. They are joining together to create a new cell. The fertilized egg cell will contain a mixture of chromosomes from the sperm cell and the egg cell.

When the fertilized egg cell divides (like the cell shown on page 26), each new cell will carry an exact copy of these chromosomes and the genes in them. The cells will continue dividing to create more and more cells. Over the next nine months, these cells will develop into a baby.

Fertilized egg cell   *Cells will keep dividing until they grow into a baby.*

Go to **www.usborne-quicklinks.com** for a link to a website where you can watch timelapse video clips of a zebrafish and a chick develop from embryos.

BACTERIA

Everywhere on Earth – in soil, air, water, plants, animals and people – are tiny living things called bacteria. They are too small to be seen with the naked eye, so until the microscope was invented, nobody knew they existed. Many bacteria are useful, but some can cause disease.

These bacteria have whip-like arms which help them to move around.

About bacteria

A bacterium is made up of only one unit or cell. Each type has a different shape, such as rounded or rod-shaped. Most bacteria group together in colonies. These sometimes form as clusters, or chains like the one on the left.

Most bacteria are so small that you could fit about 1,000 middle-sized ones on this full-stop.

These bacteria are shown magnified 48,000 times.

Two by two

The picture below shows how bacteria reproduce (make new bacteria) by dividing into two. Some divide as quickly as once every 20 minutes.

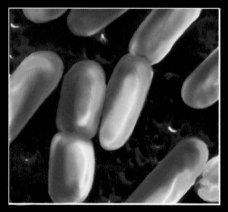

Electron microscope image of bacteria called E. coli, which cause food poisoning. They are dividing to form new bacteria.

Bacteria colonies

As long as bacteria have a supply of food to live on, colonies form and grow very quickly. Bacteria colonies thrive in warm, damp places, such as the human body, or on food that has been left uncovered.

Microscopic detail

All the bacteria on these pages were viewed with an electron microscope, so you can see the individual bacteria that make up a colony.

Jelly dishes

Scientists grow bacteria on dishes of jelly. Here is a recipe for jelly you can use to grow bacteria. You will need twelve plastic petri dishes, a vegetable stock cube, ½ a teaspoon of sugar, a small sachet of gelatin and 250ml of water.

1. Wash your hands before you start. This reduces the number of bacteria that might spread to the jelly.

2. Boil the water in a pan. Stir in the stock cube, sugar and all of the gelatin. Simmer for 30 minutes.

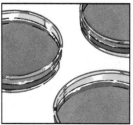

3. Pour the mixture into a jug, then pour a little jelly into each dish. Put the lids on and let the jelly set.

4. Keep the dishes upside-down until you need them. This stops water from forming on the jelly's surface.

Go to **www.usborne-quicklinks.com** for links to the following websites:
Website 1 Browse an image library with a huge variety of close-up pictures of bacteria.
Website 2 Examine the structure of a bacterial cell with the help of a clickable diagram.

Fingertip test

This experiment will show you that bacteria flourish on unwashed hands.

You will need two dishes of jelly (see page 32). Try the experiment with unwashed fingertips, and fingertips washed with soap and water.

1. Press an unwashed fingertip onto one jelly surface. Use a different dish for a washed fingertip.

2. Tape the lids onto the dishes. Label them and leave them at room temperature for a few days.

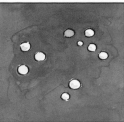

3. Without taking the lids off, place each dish in turn onto your microscope stage. Light it from above.

On the dirty fingertip jelly, you might see blobs in the shape of your fingertip. These are bacteria colonies. You should find fewer blobs on the other jelly. Never take the lids off the dishes – the bacteria could make you ill. Seal the dishes in a bag, throw them away and wash your hands.

⚠️ Caution

Useful bacteria

Some bacteria are harmless, some are even useful. Your body, especially your gut, contains various useful bacteria.

These useful bacteria live in your gut.

Body protection

The bacteria above are called lactobacilli. They help to digest your food and protect you against harmful bacteria. Another type of bacteria make vitamin K, a chemical that helps your blood to clot.

Spreading disease

Harmful bacteria can also breed in your body. They make poisonous substances called toxins, which sometimes make you ill.

The chain on the right is made up of Streptococcus bacteria. These can cause unpleasant illnesses such as sore throats, earache and skin infections. Fortunately these bacteria can be destroyed with a medicine called penicillin (see page 49).

The cluster below is made up of Staphylococcus bacteria. These cause boils and make wounds become infected.

Staphylococcus bacteria, magnified 90,000 times.

These Streptococcus bacteria (above) can cause sore throats.

VIRUSES

Viruses are the smallest living things known. They can be over a million times smaller than bacteria, and it is only possible to see them with an electron microscope. Some viruses are harmless, but many cause illnesses, ranging from the common cold to AIDS.

Flu viruses shown at over 137,000 times their real size.

Invisible germs

In the nineteenth century, scientists discovered that many diseases were caused by bacteria. But they were puzzled that they could not find the cause of diseases such as rabies or measles.

It was not until the electron microscope was invented in the 1930s, that scientists could see the viruses that cause these diseases.

Complex chemicals

Some scientists think that viruses are not living things, but just complex chemicals. They are made up of a core of genetic instructions, such as DNA (shown in blue), surrounded by a protective coat (shown in red).

Genetic core — *Protective coat*

The measles viruses at the bottom of the page are typical of many viruses.

New measles viruses being released from an infected body cell. Shown magnified over 41,000 times.

Like all viruses, measles viruses cannot live on their own. Instead, they force their way into the cells of living creatures and use these cells to make more viruses. After a time, the invaded cells die and the viruses set off to find new cells to attack.

How viruses grow

A virus breaks through the protective surrounding of a healthy cell.

The virus empties its DNA into the cell, where new copies of the virus are made.

As the cell dies, the new viruses are released into the blood, where they travel around the body to attack other healthy cells.

Changing viruses

All viruses can gradually change to form new versions of that type of virus. The viruses on the left cause flu. There are many types of flu. Each is slightly different, and scientists discover new types every year.

By using microscopes to study flu viruses, scientists have developed medicines to stop people from catching some types of flu. These medicines are called vaccines*. You can see an electron microscope picture of a vaccine below.

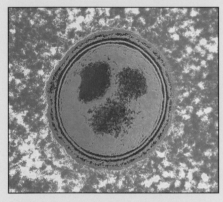

A vaccine, magnified over 60,000 times.

Cold virus

These red and yellow blobs are common cold viruses. Like other viruses, they thrive because they successfully invade other cells, especially the cells that line your nose, throat and lungs.

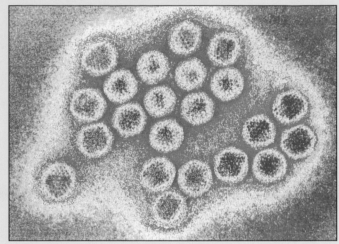

These cold viruses can be used to treat some genetic diseases.

New treatments

Scientists can now use the virus's ability to invade other cells to develop new methods of treating some diseases.

For example, cold viruses are being used to try to treat people with cystic fibrosis, a lung disease caused by faulty DNA (see pages 30 and 31).

Scientists inject healthy human genetic material into some cold viruses. Then they infect the patient with the virus, which invades the patients' lung cells. But instead of emptying their own, harmful genes into the cells, the viruses deliver healthy genes to the faulty lung cells.

AIDS

The picture on the right shows tiny HIV viruses (red) attacking a white blood cell. These viruses cause AIDS, a deadly disease that makes the body unable to protect itself from illnesses. HIV viruses are very complex and scientists are still trying to find a vaccine to prevent AIDS.

Magnification by 14,400 times shows how small HIV viruses are, compared to the white blood cells that they attack.

*Go to **www.usborne-quicklinks.com** for a link to a website with an online activity where you're a medical lab technician looking at viruses and bacteria.*

You can find out more about vaccines over the page.

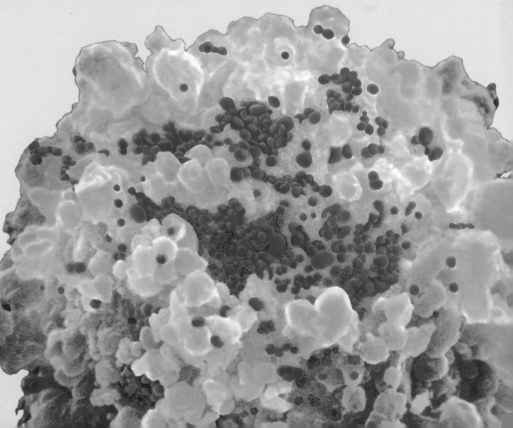

MICROSCOPES IN MEDICINE

Microscopes are used to study the way the body works and how it can be cured when something goes wrong. They have helped scientists to discover the causes of many diseases and to develop ways of treating, curing and even preventing them.

Unhealthy skin cells, magnified over 400 times.

Finding the culprit

For doctors to treat a disease effectively, they need to find out what is causing it. They can often do this by looking at blood, or samples of body tissue, with a microscope.

If the disease is caused by bacteria, doctors need to identify them. To do this, a sample, such as blood, is placed in some nourishing liquid, which allows any bacteria in the sample to grow.

This liquid is then spread thinly in streaks on a dish of clean jelly. After a few days, scientists look at the dish with a microscope to identify any bacteria that have grown on the jelly.

Scientist spreading a bacteria sample on a jelly dish.

Faulty cells

The picture above is an optical microscope view of a tiny piece of skin.

The green patches are healthy skin cells, but the red cells have become faulty. They may form a growth called a tumour which could spread and destroy the healthy cells. This disease, called cancer, can start anywhere in the body.

These cancer cells are a different shape from the healthy cells.

Healthy cell

Dyed cells

The skin cells at the top of the page have been dyed to make any cancer cells in the sample show up clearly under a microscope. If cancer is discovered early enough, it can often be treated.

Killing bacteria

Most bacteria can be killed by medicines called antibiotics. Scientists have developed different antibiotics to destroy different types of bacteria.

Some antibiotics are tested by adding them to bacteria colonies on plates of jelly. Scientists use microscopes to see if the colonies touched by the antibiotic are destroyed.

Streaks of bacteria

Antibiotic drug

Bacteria-free area where antibiotic has destroyed bacteria.

A vaccine will protect the body against this virus, which damages the liver.

Superbugs

Microscopes have shown that some bacteria are becoming resistant to antibiotics. This means that most antibiotics can no longer destroy them. These bacteria are known as superbugs. The streptococci bacteria in the picture below are a type of superbug.

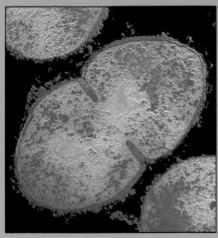

These bacteria can no longer be destroyed by ordinary antibiotics.

*Go to **www.usborne-quicklinks.com** for a link to a website where you can watch a short movie about fighting disease and try a quiz.*

Fighting disease

Microscopes have enabled scientists to discover different ways in which the body fights disease. At the bottom of the page, for example, you can see a white blood cell. It is surrounding an enemy cell, which it will then destroy.

Killer chemicals

Another type of white blood cell produces chemicals called antibodies to destroy bacteria and viruses. Antibodies recognize germs that have attacked you before. They stay in your body to stop the same germs from attacking you again.

This knowledge has helped scientists to make medicines called vaccines. These can prevent many diseases, such as measles and mumps.

Vaccines usually contain a small dose of a bacterium or virus. The dose is too weak to make you ill, but it helps your body to make the antibodies that will protect you against that illness in the future.

A white blood cell engulfing a harmful yeast cell.

Go to **www.usborne-quicklinks.com** for a link to a website where you can see some of the equipment used in microsurgery.

MICROSCOPES AND SURGERY

The surgeons pictured below are using a microscope to perform an operation. This type of surgery is known as microsurgery. It is used for operations involving minute or particularly delicate parts of the body, such as the brain, eyes or ears.

Brain work

Microscopes are essential in brain surgery because the cells which control the working of the whole body are very close together. Any damage to the brain cells could kill a patient or leave them unable to walk or talk.

Fingers and toes

Microsurgery has made it possible for surgeons to re-join parts of the body, such as fingers and toes that have been cut off in an accident.

Studying their work through a microscope, surgeons are able to re-attach the body part with a minute needle and thread. This means sewing together tiny nerves, muscles, tendons and blood vessels.

Surgeons using a microscope to help with an operation.

Joining the ends

To join together two ends of a vein that has been cut, for example, the surgeon places a metal ring around the vein. The ring holds the stitches in place while the ends of the vein are joined up.

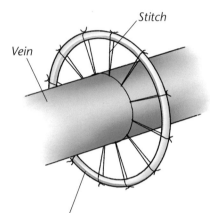

Stitch

Vein

Metal ring. This is removed when the stitches are in place.

Sharing the view

Microscopes with more than one pair of eyepieces, like the one shown here, are called multi-ocular microscopes. They are sometimes used so that other doctors and nurses can see what is happening during an operation.

Video link

Occasionally the microscope is linked to a video camera, so that student surgeons can learn about difficult operations.

PLANTS AND FUNGI

Unripe corn seeds, magnified 80 times.

LOOKING AT PLANTS

Plants, like animals, are made up of thousands of individual units called cells. But plant cells are different from animal cells. Here you can discover what some of these differences are, and take a first look at plants in microscopic detail.

Cells from a red onion, magnified 900 times. These cells need water, so they have gone floppy.

Cell patterns

The cells on the right are from an onion. They fit together in a regular, patchwork pattern.

Cell parts

Surrounding each cell you can see a dark, rigid layer called the cell wall. Inside the cell wall is a space, called a vacuole, filled with watery fluid. Both these parts help plant cells to keep their shape.

The dark spot you can see in most of the cells is called the nucleus. It controls how the cell grows and works.

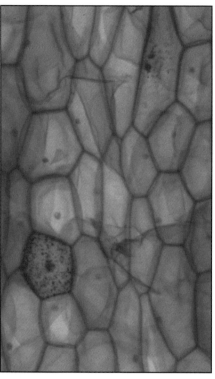

These onion cells have been stained pink to show the details more clearly.

Floppy cells

Above you can see some more onion cells. These cells do not have enough water inside them. The vacuoles within the cell walls, coloured red, have shrunk, making the cells floppy. This is why a plant droops if no one waters it.

Animal cells do not have these large vacuoles or thick cell walls. This is because they need to be flexible so the body can move around.

Cell wall   Vacuole

Plant cell   Animal cell

Looking at cells

An onion is an easy plant to collect cells from. You need to peel the brown skin away from the outside of the onion. Using a sharp knife or scalpel, carefully cut the onion in half from top to bottom. Next, cut one half in two, again from top to bottom.

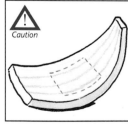

⚠ Caution

1. Take one layer of the onion flesh and cut out a piece as shown above.

2. On the inside of this piece is a thin skin, or membrane. Use tweezers to peel the membrane away.

3. Make a wet mount as described on page 45. Look at the slide, lighting it from below.

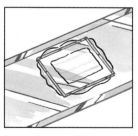

4. You may see a small, roundish patch in each cell. This is the nucleus.

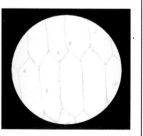

Go to **www.usborne-quicklinks.com** *for links to the following websites:*
Website 1 *Zoom in on a virtual plant cell which you can dissect to learn more about its structure.*
Website 2 *Look at fascinating microscopic images of plants, from blades of grass to pollen grains.*

1. Place the specimen in a small dish of stain and leave it there for about three minutes.

2. Use tweezers to take the sample out of the dish. Rinse it well in a bowl of clean water.

3. Make a wet mount (see page 45) and look at your stained section, lighting it from below.

WARNING: Stains can be harmful.
▲ Keep all stains away from naked flames and out of reach of young children.
▲ Always wear disposable gloves when using stains.
▲ Never try to taste the stains.

Leaf cells

The green shapes pictured below are cells that make up the top surface of a leaf. A plant breathes and makes most of its food in its leaves.

These cells fit together like jigsaw pieces to provide a protective layer. The cells look fuzzy because they are covered with a waxy, waterproof substance which stops the leaf from losing too much water.

A magnified view of the surface of a leaf.

The long, narrow holes between some cells open and close to allow gases to travel between the air and the leaf. Leaves mix gases from the air with water to make food.

Food processors

The picture below shows a cross-section of a turnip leaf. In the top half, which would be nearest to the sun, are some long cells called palisade cells. They contain a large amount of a chemical called chlorophyll. This traps energy from the sun which is used by these cells for making food.

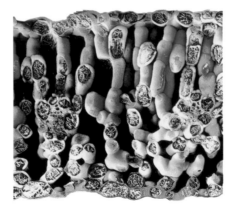

Electron microscope image of the inside of a turnip leaf, magnified 110 times.

Near the bottom of the leaf, the cells are more rounded. They help the leaf to keep its shape. In the spaces between the cells, gases such as oxygen are able to move around the leaf.

Ouch!

This sharp needle is a stinging hair on the underside of a nettle leaf. Like animals, many plants have developed ways of protecting themselves from their enemies.

When an animal or person brushes against the nettle, the pointed tip of the hair pierces the skin and breaks off. The hair injects chemicals into the victim, causing a painful sting.

This vicious-looking nettle sting is shown at 70 times its real size.

This group of cells holds the sting firm on the leaf.

Go to www.usborne-quicklinks.com for a link to a website where you can compare microscope images of leaves and find tips on how to use nail polish to make your own leaf casts.

PLANT FOOD

Like any living thing, plants need food and water to keep them alive. Specialized cells inside their roots and stems transport these vital ingredients around. The microscope pictures on these pages show the different parts of plants that do this job.

Roots

The optical microscope picture below shows the tip of a root. The root anchors the plant firmly in the ground, and collects water from the soil for the plant to use.

Protective tip

The dark, pink area is called the root cap. It protects the root by producing a layer of slime called mucus. This helps the root to push smoothly through the soil as it grows.

This cross-section of a root has been stained to show the details. The tiny red spots in the cells are nuclei, the cells' control centres.

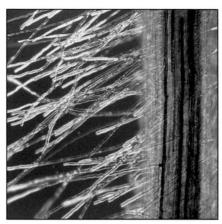

These root hairs absorb water, which helps the plant to grow.

Above you can see a close-up of the root of a beetroot. It has tiny hairs along its sides. These are where water is absorbed from the soil.

Growing sideways

The water that comes in through the root hairs collects in hollow tubes in the roots called xylem. These carry the water from the roots and stem into the leaves.

You can see xylem in the root at the bottom of the page. You can also see a smaller root growing out of the side of the main root. This helps to hold the plant firmly in the soil.

The roots in the pictures below have been cut into thin slices, or sections. (You can find out more about cutting sections on page 90.) This lets light shine through and shows details more clearly.

Root cross-section, magnified 22 times.

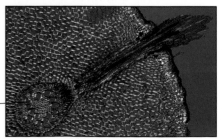

Xylem——

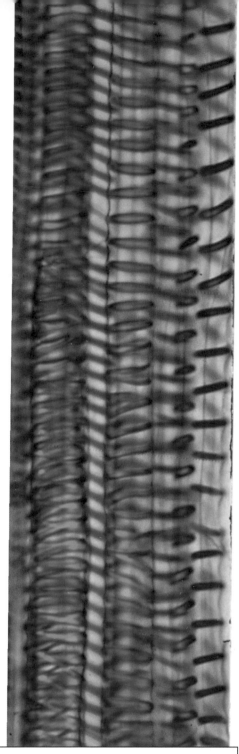

This magnified cross-section of a willow tree stem has been coloured on a computer to show the different cells more clearly.

(1) These cells form tubes which carry food from the leaves to other parts of the plant, and are called phloem.

(2) These cells form tubes which carry water from the roots upward, and are called xylem.

(3) The tissue that divides xylem and phloem is called cambium. This is where new xylem and phloem cells are made.

Transport system

The optical microscope image above shows a cross-section of a willow tree stem. Here you can see some of the cells that transport food and water around the plant. By counting the rings of xylem you can also find out how old a tree is. This tree is two years old.

Water pipes

On the far right is a section of a palm plant stem. Water travels from the roots to the rest of the plant through the long tubes of xylem cells, stained blue.

Shape keepers

The spiral bands inside the tubes strengthen the stem and help to keep the xylem tubes in shape. As the plant grows, the spiral bands stretch.

The xylem cells on the right of the picture grew when the plant was shooting up. They have stretched as the stem has grown larger. The xylem cells on the left grew later, when the plant was almost full size. You can see that they have not stretched quite as much.

Optical microscope image of part of a palm plant stem. It has been magnified over 1,900 times.

Pulling stains

Some samples, such as plant sections, are too delicate to stain with the dip and rinse method (page 41). In such cases you can use another method, called pulling. Before you start, you will need to make a wet mount of your section (see page 45).

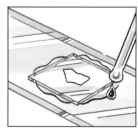

1. Use a glass rod to place a drop of stain in the water next to one edge of the cover slip.

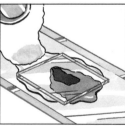

2. Hold a piece of blotting paper in the water close to the opposite edge of the cover slip.

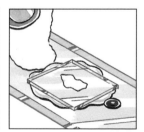

3. The blotting paper will pull the stain across. You could now add another stain in the same way.

WARNING: Stains can be harmful.
▲ Keep all stains away from naked flames and out of reach of young children.
▲ Always wear disposable gloves when using stains.
▲ Never try to taste the stains.

Go to **www.usborne-quicklinks.com** for a link to a website where you can watch a micro-video clip showing chloroplasts as they circulate around the inside of a plant cell.

HOW PLANTS SPREAD

Flowers contain parts that are used to make new plants. They make sure that seeds spread, and that new plants grow every year. This process is known as reproduction. These pictures show some microscopic views of the parts of flowers involved in reproduction.

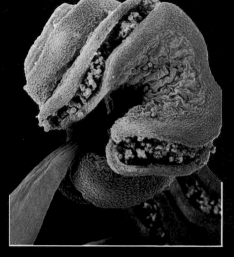

Stamen, or male part, of a rose.

Flower parts

The chickweed flower shown below has been magnified about six times. In it you can clearly see the features found in most flowers.

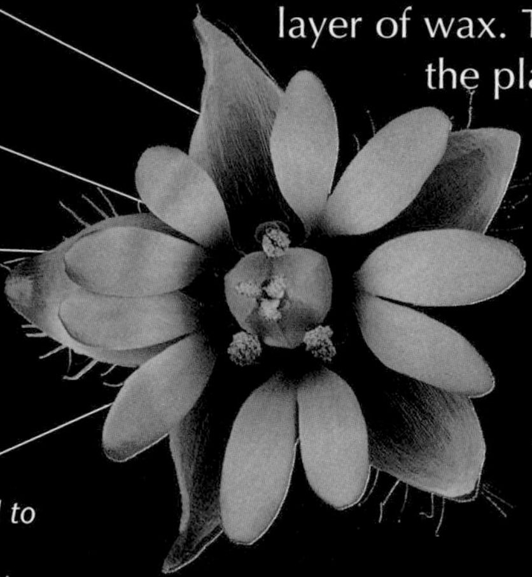

1. Leaf-like sepals protect the flower when it is still a bud. This chickweed flower has five sepals.

2. Petals are often scented and brightly coloured to attract insects.

3. These are stamens. Pollen, a yellow powder which is carried to other flowers by insects or the wind, is made here.

4. These tufts are called stigmas. Pollen grains need to land on top of the stigmas before new plants can grow.

Pretty petals

The orange, hill-like shapes at the bottom of the page show the surface of a rose petal, viewed with an electron microscope.

Each of the swellings is covered with a protective layer of wax. This prevents the plant from losing too much water through its petals.

The petal's bright colour, and its fragrant smell, attract insects.

Pollen makers

Almost every plant has male parts, called stamens. The picture above shows a rose stamen. The orange shape at the top of the green stalk is an anther. This is where pollen, a yellow powder, is made.

You can see some grains of pollen emerging from a slit in the anther. Pollen contains reproductive cells which need to be carried to other flowers so new plants can grow.

A grain of pollen

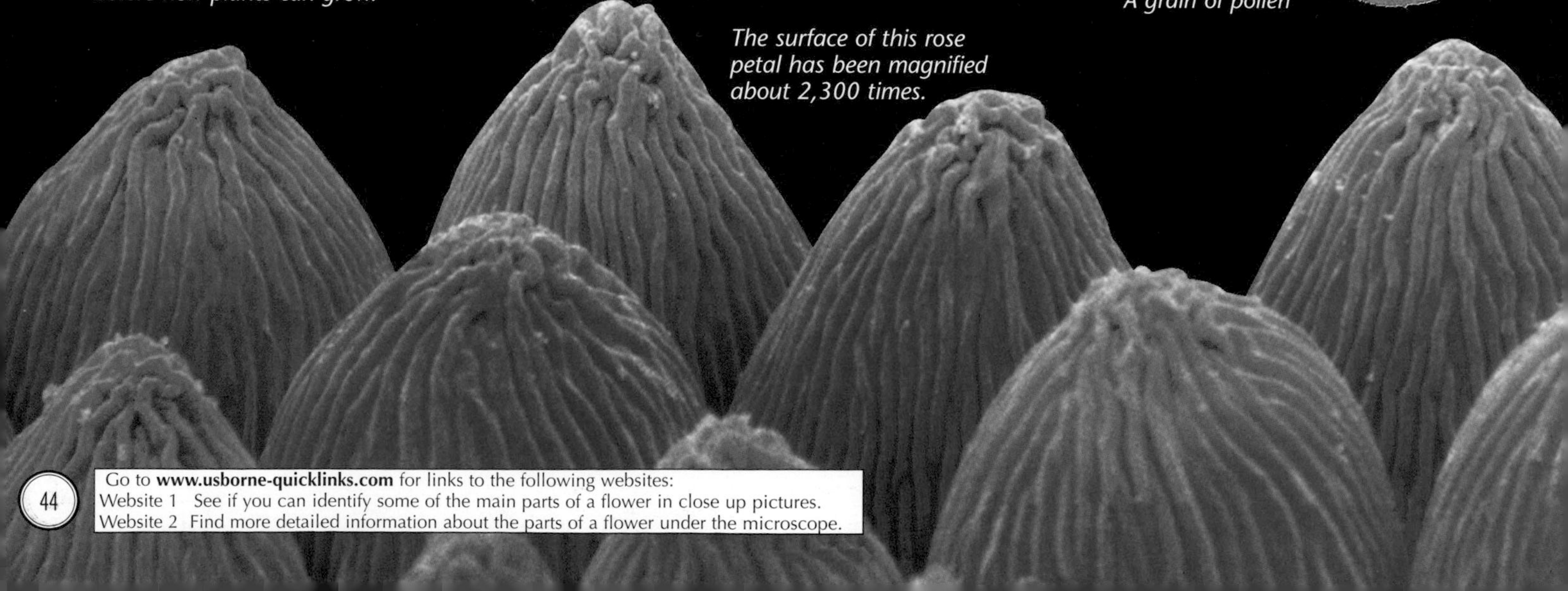

The surface of this rose petal has been magnified about 2,300 times.

Go to **www.usborne-quicklinks.com** for links to the following websites:
Website 1 See if you can identify some of the main parts of a flower in close up pictures.
Website 2 Find more detailed information about the parts of a flower under the microscope.

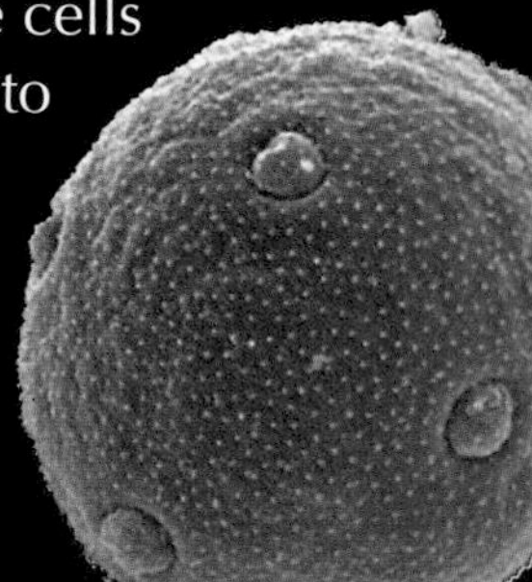

Pollen shapes

Pollen is carried from one plant to another in a variety of ways. For example, the spiky pollen grain on the right will stick to the furry body of an insect, such as a bee. When the insect flies to the next flower, this pollen rubs onto the flower's female parts, or stigmas.

Spiky pollen is carried by insects.

This rounded grain of pollen has been magnified 1,400 times.

Rounded grains of pollen are carried to other flowers by the wind. The pollen on the left has wing-like pockets which help it to float on the lightest breeze. In the summer, the air is full of pollen from trees and flowers. It gives some people hay fever, making them sneeze.

New plants

If pollen lands on the stigma, a tube grows from it into the ovary, the part of the plant which contains eggs.

Stigma
Pollen
Pollen tube
Ovary
Eggs

Reproductive cells from the pollen travel down the tubes and join the eggs to make seeds. These seeds will scatter, and grow into new plants.

Electron microscope view of grains of poppy pollen, with tubes growing down through the stigma of a poppy flower.

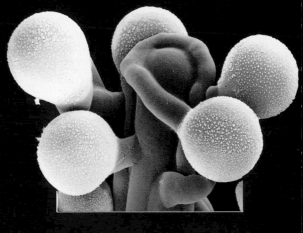

Wet mounts

Most specimens of body cells and plant parts need to be kept moist to prevent them from drying out.

To stop a specimen from drying out while you look at it with your microscope, you can make a wet, or temporary, mount.

1. Use a dropper or glass rod to place a drop of water onto a clean microscope slide.

2. Put the specimen into the drop of water with tweezers.

3. Pick up a cover slip by its edges and place one edge on the slide. Lower the slip onto the slide.

4. Use blotting paper to soak up any water which seeps onto the slide. Try not to trap air bubbles.

Go to www.usborne-quicklinks.com for a link to a website where you can find out the best way to collect, prepare and observe pollen grains under a microscope.

WATER PLANTS

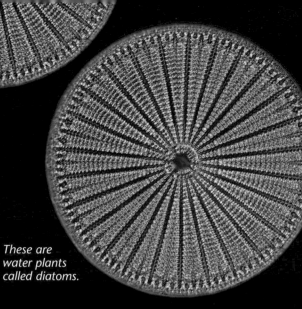

All areas of water, from small, freshwater ponds to large, salty oceans, contain a wide variety of tiny plants. Without them, nothing else in the water could survive. Here are some of the minute plants you can see in a drop of water with a microscope.

These are water plants called diatoms.

Shades of green

These beautiful green shapes are microscopic water plants called desmids. Unlike plants that live on land, tiny water plants have no roots, leaves or flowers.

Two water plants called desmids, shown at 320 times their real size. They will soon separate from each other.

Most desmids have two identical halves. New desmids are made when the two halves separate.

Green water plants, or algae, can be found in both fresh and salty water. They contain chlorophyll, a chemical that is also found in land plants. This gives the plants their green colour and helps them to use the sun's energy to grow.

Fully-grown desmid with identical halves.

A new half grows from each old half. (Same stage as big picture above.)

Desmids separate.

New halves will grow to match old halves.

Water jewels

The wheel shapes above are diatoms, a common type of water plant. They have a hard outer casing made from a substance called silica. The casing has two halves which fit together like a box and lid.

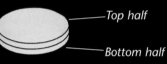

Top half

Bottom half

Under an optical microscope, diatoms sometimes look like sparkling jewels. This is because light breaks up as it shines through their glassy casings. This can create spectacular effects.

These diatoms have been magnified 160 times. No two diatoms are exactly alike.

*Go to **www.usborne-quicklinks.com** for links to the following websites:*
Website 1 *Browse a microscope image library to see amazing pictures of algae and diatoms.*
Website 2 *Watch a video of a microscopic water plant as it swims.*

Strings of cells

Some water plants are made from strings of single cells that have joined end to end. For example, the optical microscope image on the right shows a type of water life called blue-green algae.

The chains of algae are surrounded by a protective coating of jelly.

This single alga will join with other algae to form a chain.

Scientists now think that blue-green algae are not plants, but bacteria (see pages 32-33). Certain types of these algae are harmful, and release poisons into the water. If fish or other animals drink this polluted water, they can die. These poisons can also be dangerous to people.

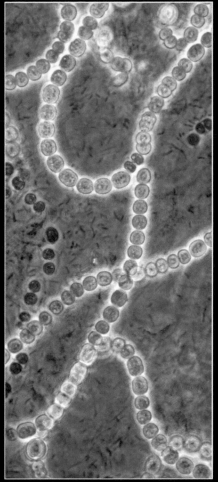
The blue-green algae above are shown about 1,000 times larger than their real size.

These feather-shaped plants are a type of diatom (see previous page).

Water testing

Scientists use microscopes to test samples of river and lake water for pollution. The number and type of plants that they can see, especially blue-green algae, help them to measure how polluted the water is.

Dark backgrounds

Most of the plants on these pages have been viewed with a type of lighting called dark ground illumination. This method uses a ring filter, a part which is attached above the light on a microscope. The ring stops light from shining directly up through the plants and makes them show up as bright objects on a dark background.

Water samples

You can collect water samples in clean jars. Collect fresh water from ponds or lakes, and salt water from the sea or rock pools. You may also find water life on slimy pebbles. Label the jars to show where you collected the samples.

1. Pour some of the water into a dish and leave the specks in it to settle for a few minutes.

2. Use a dropper to take up a small amount of water that contains some specks.

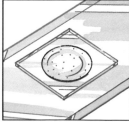
3. Place a few drops of the water onto a cavity or ring slide* and put a cover slip over the top.

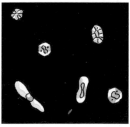
4. Light it from above. If you have a ring filter, try using dark ground illumination (see *Dark backgrounds*, above).

Go to **www.usborne-quicklinks.com** for a link to a website where you can view microscopic pictures of marine diatoms from around the world and find out more about them.

*See equipment list on page 88.

FUNGI

ungi can be seen everywhere. They appear as mould on rotting food, mildew on musty walls, or mushrooms in a field. In many ways they are quite unlike plants and animals. Here you can see what these curious organisms are made of.

The mould on this orange is a type of fungus.

Looking at fungi

In the picture on the right you can see the two main parts of a typical fungus. The red blobs are called fruiting bodies and the pale, thread-like tubes are called hyphae.

Tangled mesh

The hyphae are loosely tangled together into a mesh called mycelium. This threads its way under the surface the fruiting body grows on.

This mushroom mycelium forms a complex web underground.

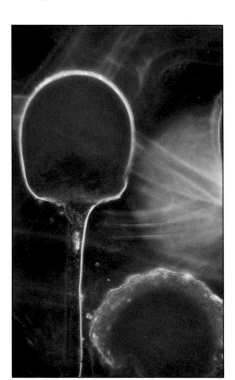

Fungi under an optical microscope. The red beads will burst and scatter spores which will grow into more fungi.

Fruiting bodies

Fruiting bodies are made up of very tightly-packed hyphae. The ones in this picture have been magnified over 1,100 times, so you can imagine how small they really are.

Scattering spores

Inside each fruiting body are thousands of microscopic dust-like seeds called spores. When the spores are ripe, they will burst out of the fruiting body and be scattered by the wind. If the spores land on suitable food, new fungi will start to grow.

Spore prints

Fungus spores are very easy to collect and interesting to look at under a microscope. Experts often look at spores to help them identify fungi. The best way to collect spores to look at is to make a spore print on a plain microscope slide.

Caution

1. Cut the cap off a mushroom and place it over a slide. Cover it with a bowl and leave it overnight.

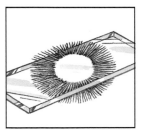

2. When you remove the bowl and the mushroom cap, you will see a spore print.

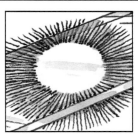

3. Do not put a cover slip over the spores, as this could damage them. Light the spores from below.

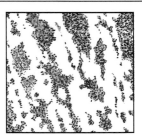

4. With a low-power lens, the spores look like streaks of specks. A high-power lens will show their shape.

Growing mould

You can grow some fungi yourself. If you use the method shown here, the two types of moulds which you are most likely to see are called Mucor and Penicillium. Mucor is a white mould with small black dots. Penicillium is blue.

1. Dampen a small piece of bread and put it in a bowl. Leave it open to the air for a few hours.

2. Cover the bowl with a lid or some clear plastic film. Keep it in a warm place for 3 days.

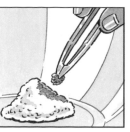

3. Use tweezers to pull off some mould. Make a wet mount (see page 45) and light it from below.

You should see the hyphae, mycelium, and fruiting bodies. Afterwards, seal the slide and bread in a plastic bag and throw it away. **Caution** Keep mould away from your mouth and nose, as breathing in spores can harm you. Clean the bowl with disinfectant, then wash your hands well.

Friends and foes

The mould you can see in the picture below is a useful fungus called Penicillium. It is used to make the antibiotic penicillin, which kills many types of bacteria.

A magnified view of a type of Penicillium mould growing on a dish of jelly. The white area at the edge is mycelium.

There are also some harmful fungi that can cause diseases. These can be mild complaints such as athlete's foot in humans, or serious problems such as Dutch elm disease in elm trees.

WARNING: Wash your hands when you have been touching fungi. Never put fungi near your mouth. They might be poisonous.

Why food rots

The large, flower-like picture below shows the fruiting body of Mucor, a common type of mould that grows on bread.

Breakdown

As the mould feeds on the bread, it releases chemicals onto the surface of the bread. These chemicals make the bread decompose, or rot.

Mouldy patches

If you are hungry and the last slice of bread is looking mouldy, don't be tempted just to pull off the patches of mould. These patches are made up of fruiting bodies, like this one.

Electron microscope image showing a fruiting body of bread mould, magnified 1,100 times. It has been coloured yellow by a computer.

What you cannot see are the delicate mycelium threads that have already spread throughout the bread. It is not safe to eat food that has gone mouldy, because some fungi can be poisonous and make you ill.

Go to **www.usborne-quicklinks.com** for links to the following websites:
Website 1 Have a look at microscopic images of fungi and mould at an online image library.
Website 2 Discover fascinating facts about fungi and view close-up images of mushroom gills, pores and spores.

FOOD SCIENCE

Microbes such as fungi and bacteria can be harmful if they infect food. But some microbes can actually be used to make food too.

These Clostridium botulinum bacteria, magnified 60,000 times, thrive in badly-preserved canned meat. They cause a deadly disease which destroys the heart or lungs.

Microbe menu

The red stripes in the first picture below, for example, are a fungus called Penicillium roqueforti. This fungus is an important ingredient in blue cheeses. The second picture shows a type of bacteria which is used to make natural yogurt.

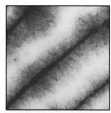

These "veins" from a blue cheese are caused by a fungus. The picture was taken through a coloured filter.

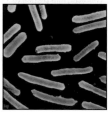

Yogurt is made by adding a type of bacterium called lactobacilli to warm milk.

Microbe menace

The picture below shows a harmful bacterium called Salmonella. It can infect food that is left uncovered, or kept for too long. Eating food that is contaminated by harmful microbes can make you ill.

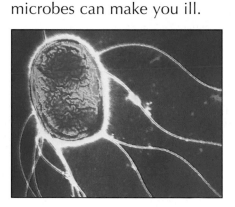

A harmful Salmonella bacterium, shown here magnified about 21,000 times.

Preventing microbes

Food manufacturers use a variety of methods to prevent harmful microbes from infecting food. For example, heating or cooling a product will affect the number of bacteria that can grow in it.

Boiling liquids at high temperatures destroys microbes.

Freezing food makes it too cold for microbes to grow quickly.

133°C

-20°C

Scientists use microscopes to check food that has been boiled or frozen to make sure that it is safe to eat.

Food test

All foods contain harmless bacteria. This test can be used to look for bacteria in flour. You can also try testing milk or yogurt.

WARNING:
▲ Never use meat to test for bacteria.
▲ Always be careful with boiling water.

1. Mix 250ml of cooled, boiled water into 10g of flour. (Boiling kills any bacteria in the water.)

2. Use a boiled and cooled teaspoon to put drops of the mixture on a jelly-filled petri dish (see page 32).

3. Tape the lid onto the dish, label it and leave it in a warm place for a few days.

Put the unopened dish under the microscope to look at the jelly. Don't take the lid off the dish. If the flour contains any bacteria, blobs of them will form on the jelly.

When you have finished, seal the petri dish in a plastic bag, throw it away and wash your hands well.

Go to **www.usborne-quicklinks.com** for links to the following websites:
Website 1 *Take a close look at food and see a surprising array of different structures.*
Website 2 *Learn about the microbes in snacks and how they help to preserve foods.*

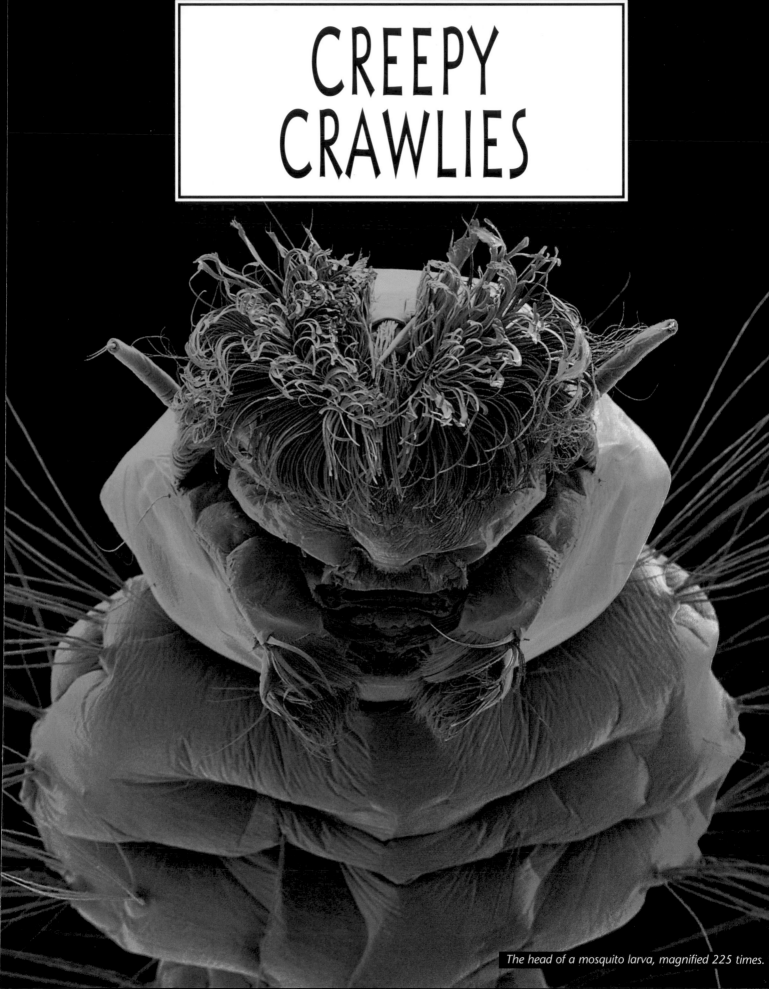

CREEPY CRAWLIES

The head of a mosquito larva, magnified 225 times.

INSECT WATCHING

There are well over a million different kinds of insects. Altogether, they outnumber people by 200 million to one. Seeing insects under a microscope helps us to understand how their extraordinary bodies have enabled them to become the most successful animals on the planet.

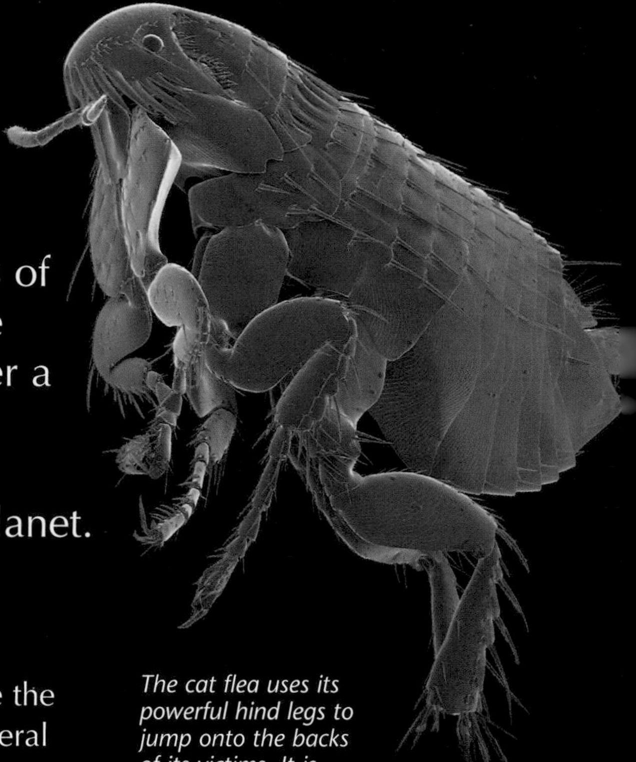

Bug bits

All insects have six legs, and a body which is divided into three sections – the head, thorax and abdomen. In this electron microscope picture of a black fly you can see the head and thorax, and several of the legs. Like this fly, most insects have wings, but a few types do not.

The cat flea uses its powerful hind legs to jump onto the backs of its victims. It is shown here 85 times larger than its real size.

Actual size of cat flea ➤

Insect bodies have three distinct sections.

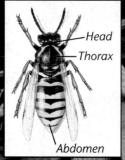

— Head
— Thorax

Abdomen

Small is great

Because it is so small, a flea like the one above can survive on minute amounts of food. It feeds on blood which it sucks from its prey with its sharp, piercing mouthparts. Its tall, narrow shape helps it to run swiftly between the hairs on the body of its victim.

Dead easy

Live insects are difficult to look at under a microscope because they move around too much. It is much easier to look at dead insects. These are easy enough to find, so don't kill insects just to look at them.

This picture of a black fly is shown over 50 times larger than life-size.

Bit by bit

This false-colour electron microscope picture of a malaria mosquito shows all the features of a typical insect.

The head

As with almost all creatures, this is where the eating and seeing goes on.

The mosquito's eyes are coloured green in this picture. All insects have eyes that are made up of hundreds of little lenses. This kind of eye is called a compound eye.

Attached to the head are the antennae, which help the mosquito sense what is going on around it. Insects use their antennae to feel things, pick up vibrations, and smell.

A mosquito feeds on blood, which it sucks out of its victim. It does this with a sharp, piercing tube called a proboscis, which you can see here coloured in red.

The thorax

The legs and wings are attached to the thorax, which is packed with muscles.

Insect wings are made of strong, lightweight material. Being able to fly is one of the main reasons why insects have become so successful.

Mosquitoes live on water. Their spindly legs help them to spread their weight over it, and walk on the surface without sinking.

The abdomen

This is where an insect digests its food and where its reproductive parts are.

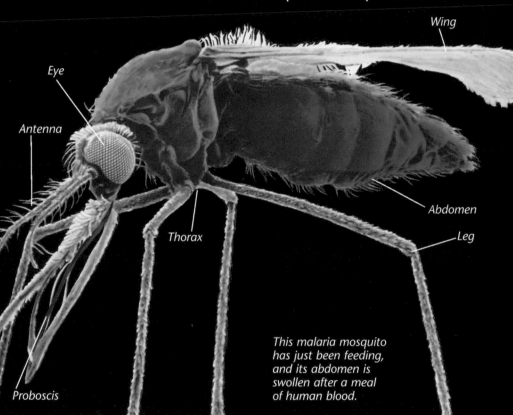

Wing

Eye

Antenna

Abdomen

Thorax

Leg

Proboscis

This malaria mosquito has just been feeding, and its abdomen is swollen after a meal of human blood.

Watching insects

Large, live insects are difficult to examine under a microscope. It is hard to keep them in one position, so it is difficult to focus on them. Small insects, such as ants, are easier to look at. Here is how to make a slide trap so you can look at them.

1. Cut a piece of thick cardboard the same size as a slide. (Draw around the slide to get the right size.)

Caution

2. Cut a slot in the middle with a scalpel or a craft knife. Keep the piece of cardboard that is cut out.

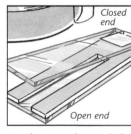

Closed end

Open end

3. Place a clean slide on either side of the cardboard. Then fasten the closed end with adhesive tape.

4. Use the cutout section of cardboard to trap your insect. View with low power, lighting it from above.

Go to **www.usborne-quicklinks.com** for links to the following websites:
Website 1 *Browse a gallery with over a hundred microscopic images of insects.*
Website 2 *Get in close and examine a mosquito that can carry a deadly virus.*

INSECTS UP CLOSE

From the claws of a flea to the sting of a bee, all insects have particular body parts which help them survive in a world where death can be an instant away. Microscopes enable us to study these body parts in minute detail, even on the tiniest insects.

Claw of a flea magnified 520 times.

Eyes

Insects have the most intricate eyes of all animals. Most have compound eyes, which are made up of hundreds or thousands of different lenses. The bulging eyes of the fruit fly in the pictures below help it to see in front, behind and above at the same time. If you have ever tried to swat a fly, you will know how effective these eyes are at spotting movement.

1. Here you can see a fly's eyes. They are placed on the side of the head and give the fly almost all-round vision.

2. Up closer, you can see individual lenses in each eye. This fruit fly has hundreds of lenses. Some insects, such as dragonflies, have 30,000 lenses in each eye.

3. Closer still and you can see miniature hairs protruding between the lenses. These help to protect the eye from microscopic dust particles.

Claws

Many insects have claws on their feet, but few look as sinister as the talons of a blood-sucking flea. The large front claw shown above grips an animal's fur, and the smaller spikes help stop the flea slipping while it feeds.

The pincers below belong to an earwig. They are used for defence and to capture prey.

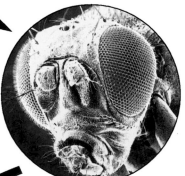

This earwig's pincers are attached to its rear, rather than its front.

Preparing insects

Insects are often too big to look at whole, but you can take the parts off a dead insect to look at. First you will need to soften the body by soaking it in 100ml of boiling water mixed with 100g of washing soda, which you can buy in a supermarket.

1. Pour the water into a jar and stir in the washing soda. Put the insect in, and leave it for three or four days.

2. Remove the insect with tweezers, and hold it under slowly-running water to rinse off the solution.

3. Hold the insect in a pair of tweezers. Pull off the part you want to view with another pair of tweezers.

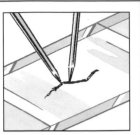

4. Place the part on a slide using mounting needles*. If possible, cover it very carefully with a cover slip.

Winged wonders

The shimmering colours you can see in the picture at the bottom of the page are tiny overlapping scales on the wing of a Priamus butterfly.

Priamus butterfly

Many butterflies are poisonous, and bright colours act as a warning to other insects not to eat them. The colours also help butterflies to recognize a mate.

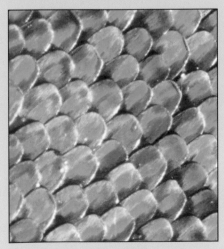

You can see individual scales on this microscopic view of a Priamus butterfly wing.

Mouth hoover

The fly in the bottom corner feeds by squirting saliva onto food. This helps to turn its meal into a sticky mess, which the fly then sucks up with its spongy mouthparts.

Insects have different types of mouths, depending on the type of food they eat.

A butterfly's mouth has a long tube for sucking juices from inside plants.

A grasshopper has sharp, scissor-like mouthparts for munching grass.

A fly sucks up liquid food with these spongy mouthparts.

Handy hints

Here are some tips on how to look at insect parts once you have removed them.

Legs. Light them from above.

Eyes. Remove the insect's head and mount it on its side on a slide.

Wings. Place on a mount and light them from below.

Butterfly and moth wings are very delicate and should not be soaked. Light them from above.

Go to **www.usborne-quicklinks.com** for a link to a website where you can discover what a lacewing looks like under a microscope and read some intriguing facts.

*See equipment list on page 89.

WATERY MINIBEASTS

From little ponds to gigantic oceans, all areas of water contain tiny animals. Some will stay at this minute size for their whole lives, but many are the newly hatched young of much larger animals. These creatures hatch in their millions, because so many are eaten before they become adults.

Mini shrimp

The beach-dwelling creature below is a miniature relative of the shrimp and lobster. It is so tiny it lives in the spaces between grains of sand. It is shown here nearly 300 times its actual size.

This red and white minibeast is shown 300 times its actual size. It is called an Antarctic rotifer.

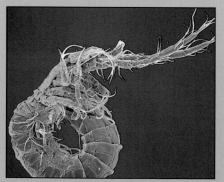

The shrimp-like creature in this picture is no bigger than a grain of sand.

Single cells

This sea-dwelling animal is called a dinoflagellate. It is so small it is made of only one cell. Its long tail keeps it stable in the water, and the two horn-like limbs at the bottom wave around, enabling it to move about.

A single-celled dinoflagellate.

Two mouths

The creature above is called a rotifer. It lives in lakes and pools in the Antarctic. At the top of its head are two sets of flailing little hairs, which waft even tinier plants and animals into its two mouths.

Although it is minute, the rotifer can reproduce in such numbers that it turns the floor of a lake bright red.

Top and tail

This microscope technique uses two microscope slides to enable you to look at both sides of a tiny water creature.

You can obtain these animals in the same way as you collect water plants (see page 47).

1. Place a few drops of water, containing the creature you want to look at, on a plain microscope slide.

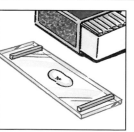

2. Gently place a piece of matchstick at each end of the slide, as shown.

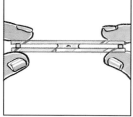

3. Place a plain microscope slide on top, so that the water is suspended between the two slides.

Light from below.

4. Tape the slides in position. Carefully flip them over to look at the underside of your specimen.

Go to **www.usborne-quicklinks.com** for links to the following websites:
Website 1 *Go virtual pond-dipping and find out more about a freshwater pond's microscopic inhabitants.*
Website 2 *Swim into the depths of a virtual ocean to discover what miniature creatures are lurking beneath the waves and find out how to collect your own.*

Death in miniature

Most microscopic animals and plants are eaten by other animals but, as these pictures show, it is not just the bigger ones who eat the smaller ones.

Here you can see a life and death struggle between a smaller Didinium and a larger Paramecium – two types of mini-creatures called protozoans.

1. As the Didinium attacks, the Paramecium fires off sticky threads, in an attempt to deter it.

2. Although the Paramecium is almost twice its size, the Didinium manages to seize it in its mouth.

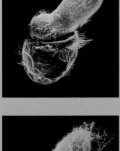

3. The Didinium struggles to twist its prey around, and then pre-pares to swallow it whole.

4. The Paramecium gradually disappears into the Didinium, to be slowly digested.

The two creatures in the sequence of photographs above are shown over 200 times larger than real life.

Tiny suckers

Some types of creatures, called parasites, do not usually kill their prey, but attach themselves to their victims and suck nourishment from them.

The ones below are fish parasites. On the right you can see two suckers near the mouthparts, which anchor the parasite to the fish as it feeds.

This microscope photograph shows two fish parasites around eight times larger than life.

Baby mosquito

Many tiny water creatures are the young of larger animals. This sinister shape is a baby mosquito. Mosquitoes lay their eggs in fresh water. They hatch into little creatures like this, called pupae, which look different from the adults.

This primitive eye can detect movement. If the pupa senses danger it sinks below the water's surface.

Breathing horn

Floating

As the pupa floats under the surface of the water it is changing and growing. After a few days it will break out of its casing, and emerge as an adult.

The pupa floats just below the water's surface, and breathes through two horns on its head.

*Go to **www.usborne-quicklinks.com** for a link to a website where you can get a close-up look at more pond life and find some interesting facts.*

UNWELCOME GUESTS

This dust mite is shown 250 times larger than life. It is perched on the tip of a needle.

Your house is infested with a countless number of tiny creatures. Many mattresses, for example, contain over a million insects, most of which can only be seen with a microscope. Many of these unwanted guests are harmless, but some can spread infection and disease.

Dust to dust

Inside your house a slow fall of dust settles on every surface. Most of this dust is flakes of human skin, which your body sheds all the time.

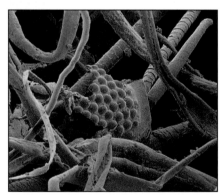

This false-colour electron microscope photograph of household dust shows skin, hair, fibres of clothing and, in the middle, the remains of an insect's eye.

Dust munchers

Wherever there is dust, there are dust mites. These microscopic relatives of spiders and scorpions feed on flakes of human skin.

Dust mites are almost impossible to destroy. Even if you vacuumed a small carpet for 20 minutes, you would only remove about 2% of all the dust mites that lurk within it.

Dust detective

You can easily collect samples of different types of dust to look at for yourself, under a microscope.

Surprisingly, the ingredients in the dust you collect will vary quite a bit, depending on where you collect it from.

1. Go to a dusty area of your house such as a window sill. Gently brush a little dust onto a slide.

2. Place a cover slip over your sample, and then examine it under a microscope, lighting it from above.

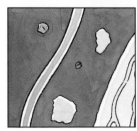

3. See if you can recognize any of the dusty debris shown on these two pages.

4. Look out for pieces of human skin, paper or clothing. If you live in a town, you may see particles of soot.

---

*Go to **www.usborne-quicklinks.com** for a link to a website where you can meet the 12 most unwelcome guests close up and study their habits.*

Egg menace

One of the most dangerous household pests is the house fly. These insects land on your food. As they eat, they leave a grubby trail of germs and eggs behind them. You can see a fly laying an egg in the picture on the right.

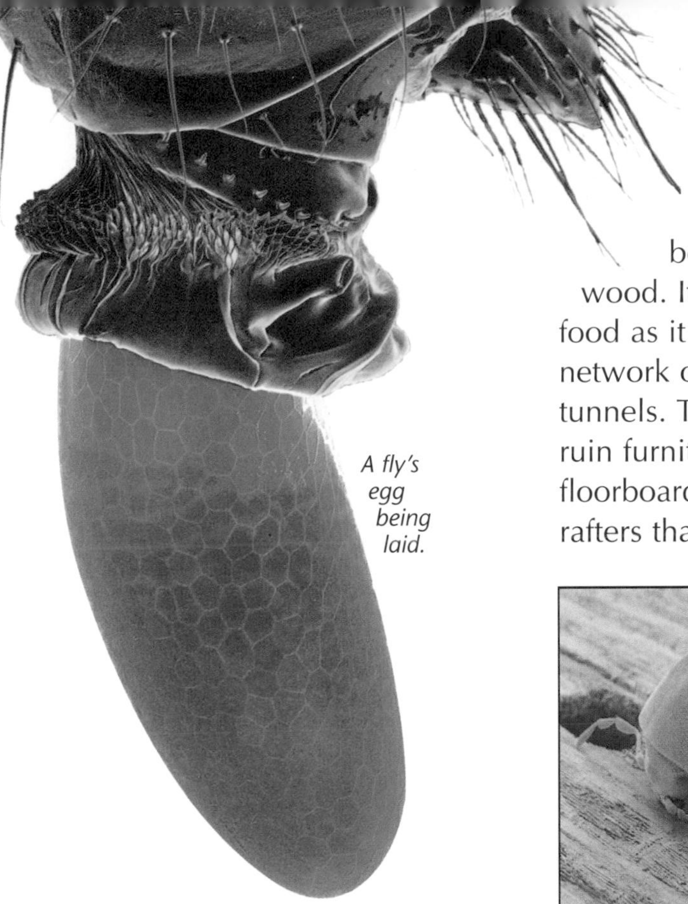

Maggot meal

Flies lay their eggs where there is plenty of food so that when the eggs hatch, the maggots that emerge can begin eating immediately.

The maggot below is eating raw beef. It feeds in the same way as an adult fly. It squirts saliva on its food to soften it, and then sucks up the soggy liquid.

A fly's egg being laid.

This fly maggot will change into an adult fly within two or three days.

This needle-like mouthpart pierces human skin and sucks out blood.

Bed biter

Repulsive bed bugs, such as this one below, live in mattresses and pillows. They use long, sharp mouthparts to pierce human skin and feed on blood. This can cause unpleasant itching in their victims.

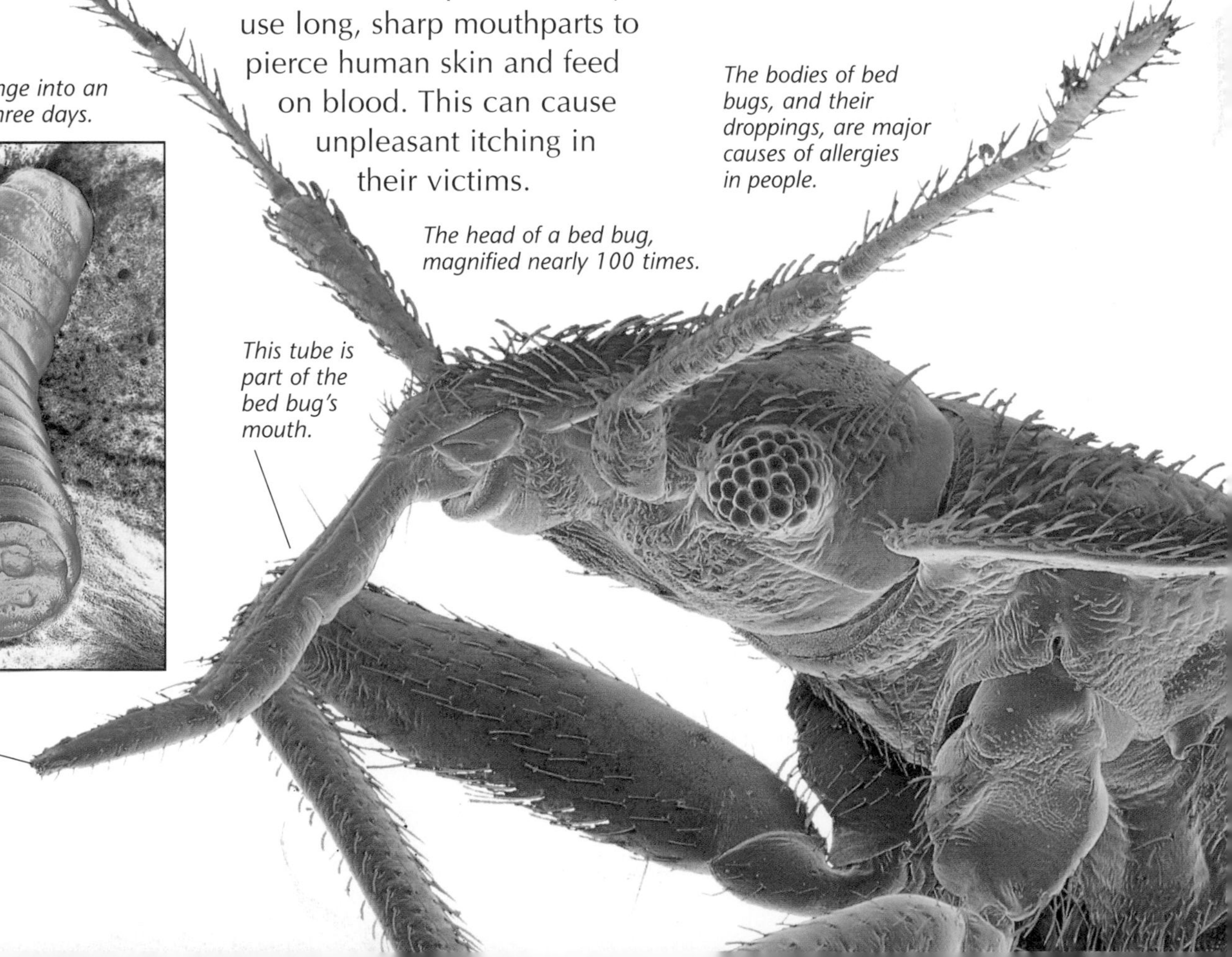

The head of a bed bug, magnified nearly 100 times.

This tube is part of the bed bug's mouth.

Wood eaters

The woodworm beetle below eats wood. It burrows into its food as it feeds, creating a network of tiny holes and tunnels. These beetles can ruin furniture, and damage floorboards and the beams and rafters that hold a house up.

The woodworm beetle in this false-colour electron microscope photograph is shown 25 times larger than in real life.

The bodies of bed bugs, and their droppings, are major causes of allergies in people.

BODY RESIDENTS

I t is not only your house that is home to a host of tiny creatures. Human bodies sometimes have minute residents too. Most are so small they can only be seen with a microscope. Some can cause illness, discomfort and even death. Here are a few of the more unpleasant varieties.

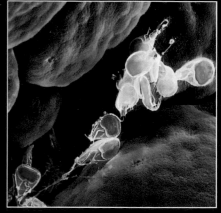

These creatures live in people's bodies.

Lice attack

These two monsters are lice, and they live on human body hair. They feed five times a day by sucking blood from their hosts, which can cause severe itching and rashes. Lice pass from one person to another through close body contact, or shared bedding and clothing.

An adult louse

Each of the louse's six legs ends in a massive claw. These grip the hair tightly whenever anyone tries to remove them.

These lice are shown nearly 70 times larger than life-size. They live in human body hair.

An infant louse

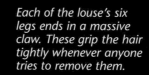

Food tube feeders

The green, pear-shaped creatures in the picture above live inside a human food tube, or intestine. They are called Giardia lamblia. They enter the human body through infected food or water.

Giardia lamblia feed on the food that passes through the human intestine. They anchor themselves to the wall of the intestine with sucker pads.

Sucker pad

Unfair exchange

The Giardia has a safe and constant supply of food. But humans who carry this creature may suffer from painful stomach cramps, diarrhoea and sickness.

Go to **www.usborne-quicklinks.com** for links to the following websites:
Website 1 Browse an image gallery for a close encounter with a variety of ticks.
Website 2 See amazing close-up pictures of lice found on humans and animals.

Skin diggers

These eggs have been laid in human skin by a scabies mite. The mite punctures a hole in the skin and burrows beneath the surface to lay its eggs. The eggs then hatch into larvae (tiny, worm-like baby mites).

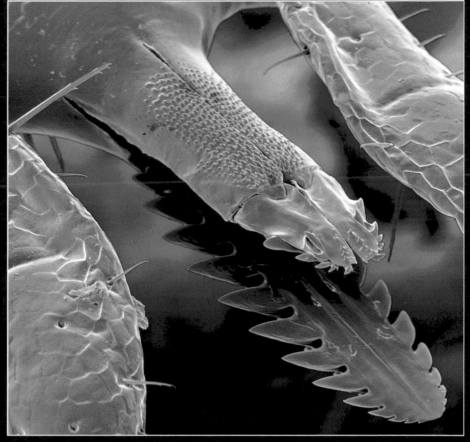

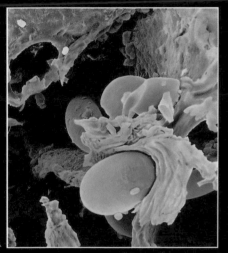

Eggs from a scabies mite, shown in human skin.

Mouthparts of a blood-sucking tick (see picture below), magnified about 350 times. The lower part drills under the skin and the upper part sucks up blood.

As the larvae burrow, the body produces liquids to try to repair the damage to the skin. The growing larvae feed on these liquids.

Tick fever

The tick on the right feeds on human blood. Most blood-sucking creatures feed on their victims for a few hours and then leave. This tick stays attached for several days. It swells up with blood and can increase its weight by up to 200 times.

Electron microscope image of a tick swollen with human blood. Such a meal can last the tick for several years.

Blood sucker

The tick's mouthparts have two sections. It uses the part at the end of the lower jaw to drill into the skin. The jagged edge anchors the drill firmly into the victim, making the tick almost impossible to dislodge. It then sucks blood through the shorter mouthpart above it.

The tick infects its victims with an illness called Lyme disease. This causes fever and painful swellings in the arms and legs.

THE PLANT EATERS

Many insects are harmless, but some can cause great damage to plants. The insects which do this are known as pests. Here is a selection of them.

Plant sucker

The little rose aphid shown below sucks the juices from plant stems and buds. If enough of these insects are present on a plant they can kill it, before moving on to their next victim.

Wood muncher

This bark beetle, seen here walking over a piece of wood, is shown 12 times larger than life.

The bark beetle above gnaws through the tough outer bark of trees to eat the softer wood beneath. This can cause serious damage to the structure of the tree.

The beetle also lays its eggs in trees. When the eggs hatch, the baby beetles tunnel farther into the wood. If enough beetles are infesting a tree, they can destroy it.

Wheat eater

The startling monster below is a baby grain weevil. It is eating its way through a grain of wheat. Its long snout ends in a razor-sharp mouth, which can chew through the tough outer casing of its food into the nutritious centre.

Adult weevils lay their eggs in grain, so that their newly hatched offspring will have a ready supply of food.

Electron microscope photograph of a grain weevil emerging from a grain of wheat. This weevil is a type of beetle.

Both crops and grain stores can be affected by grain weevils. At one time these insects caused famine, but in many parts of the world today, chemical poisons and improved storage methods have kept them under control.

This false-coloured electron microscope image shows a young rose aphid feeding on a rose leaf. It is shown over 90 times larger than its actual size.

Go to **www.usborne-quicklinks.com** for links to the following websites:
Website 1 *Take a close-up look at aphids in an online photo gallery and zoom in on their eyes, legs and antennae.*
Website 2 *Explore an online photo gallery of insects and find out where they live and what they eat.*

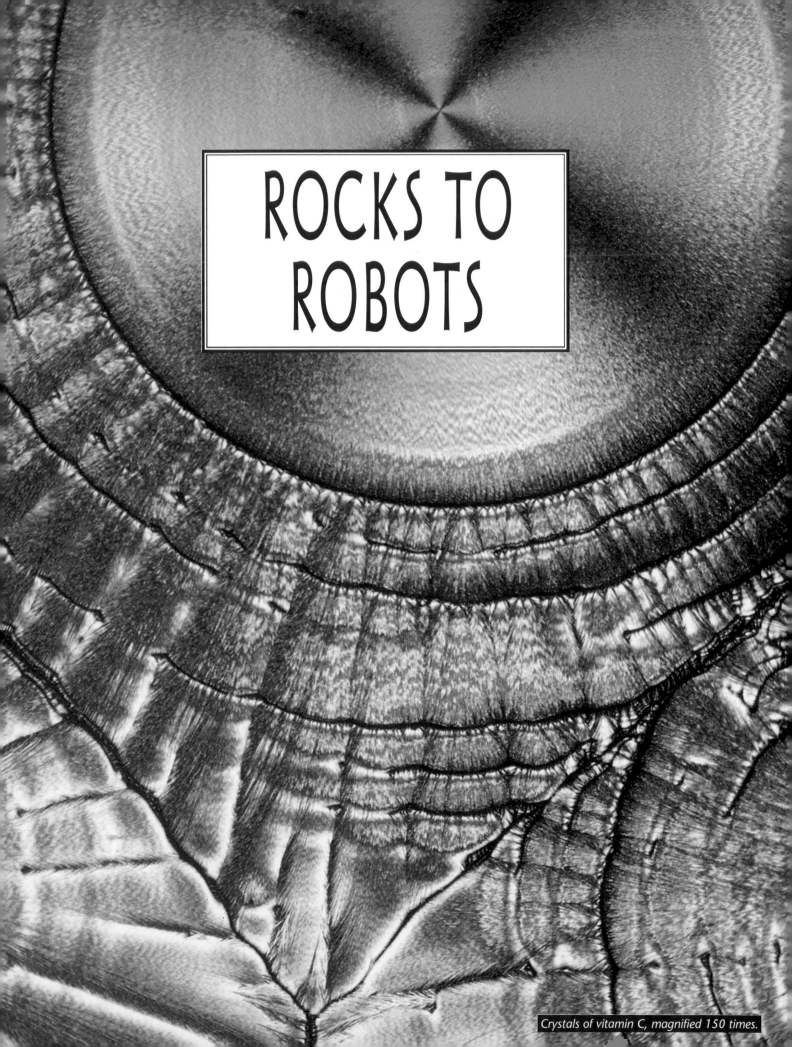

ROCKS TO ROBOTS

Crystals of vitamin C, magnified 150 times.

SAND AND ROCKS

All rocks are made from substances called minerals, which are formed inside the Earth. By examining samples of sand and rocks with microscopes, scientists can identify the minerals in them, and discover how they were formed.

Grains of sand

These rock-like objects are grains of sand that have been magnified about 100 times. Sand is made up of microscopic fragments of larger rocks that have gradually been worn away by wind, rain and water.

Most of the sand grains in the picture are tiny particles of a mineral called quartz. The shapes of the grains depend on where the sand was found.

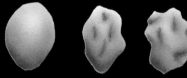

Electron microscope image of grains of sand.

Sandy shapes

Microscopes have shown that sand collected from sand dunes has smooth grains. This smoothness is caused by the wind constantly blowing the grains against each other.

Grains of beach sand have a fairly smooth surface, as they have been rubbed together by the movement of the sea.

Sand grains from a riverbed tend to be jagged and sharp. They have only recently broken from rocks, and have not had time to wear smooth.

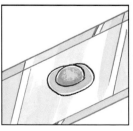

Desert sand   Beach sand   River sand

Viewing rocks

Small, flat rock pieces are best to look at because they fit easily under your objective lenses.

> **WARNING:**
> When focusing your microscope, be careful not to hit and damage the lens on the rock.

1. Wash your rock pieces in clean water to remove any dust or loose grit from the surface.

2. Place small rock samples on a glass slide, and larger ones on a piece of stiff cardboard.

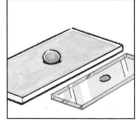

3. Hold the sample in place with modelling clay. Place the flattest surface uppermost, as it is easiest to focus on.

4. Use a dropper to wet the surface. This can help to show the detail. Light the rock from above.

Go to *www.usborne-quicklinks.com* for a link to a website where you can look at beautiful close-up images of sand as you take an unusual trip to beaches and bays around the world.

Polishing rocks

You will be able to see more detail in your rock samples if you polish their surfaces to make them smoother. Find a rock that is larger and harder than the rock you want to polish. (A harder rock will be able to scratch the surface of your rock.)

1. Wet your piece of rock and rub it against the larger rock, using a circular motion as you rub.

2. Keep wetting the rock as you rub, and look at the surface from time to time to see how smooth it is.

3. To polish your rock sample further, use fine grade emery or sand paper. Lay the paper on a flat surface.

4. Rub the rock on the paper using a circular motion. Keep wetting the rock as you polish it.

Sand to stone

Below you can see some grains of quartz sand in a cross-section of sandstone. They are held together by mud and clay, which appear here as black specks.

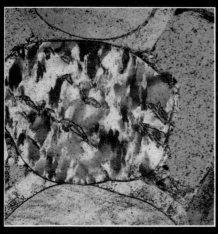

Cross-section of sandstone, magnified 120 times. The green is resin, which was injected into the rock to stop it from falling apart when the section was cut.

Sandstone is a type of rock called sedimentary rock. It forms over many years from layers of particles (such as sand) that build up on the bottom of seas or lakes. These layers become tremendously heavy and crush down deeper layers, turning them to rock.

Mineral patchwork

This picture shows a section of a rock called granite. It is made from different minerals that have joined together like patchwork. The flat-sided shapes you can see here are called crystals.

Granite is a type of rock called igneous rock. It forms when hot, liquid rock from deep in the Earth cools down. This can happen when lava erupts from a volcano and then cools.

Section of granite, magnified 16 times. It is viewed with a type of light called polarized light, which makes the minerals in the rock show up in different colours.

Go to **www.usborne-quicklinks.com** for links to the following websites:
Website 1 Explore an online exhibit about the microscopic world found inside rocks.
Website 2 An introduction to observing rocks under a microscope with close-up images of a variety of minerals.

MICROFOSSILS

Some types of rocks contain the remains of minute plants and animals that lived in the seas and oceans millions of years ago. Many of these remains can only be seen with a microscope, so they are called microfossils.

Microfossils viewed with an electron microscope.

Tiny creatures

On the right, you can see a thin section of limestone that has been viewed with polarized light*. The large, pink and brown objects are the remains of minute sea creatures called foraminifera.

Crushed skeletons

The rock which surrounds them is made from the crushed skeletons of even tinier creatures. Rock scientists, called geologists, estimate that the creatures in this rock are between 37 and 54 million years old.

Shells of microscopic sea creatures in a piece of limestone, magnified 20 times.

Turned to rock

When sea creatures such as foraminifera die, they sink to the seabed. Here they become trapped in other particles (called sediment) that have settled there. Over many millions of years this sediment gradually turns into rock.

Chemical changes

As the rock forms, great pressure from the layers of sediment above causes chemical changes in the bodies of the creatures trapped there. Their shells turn into fossils.

Finding fossils

You can look for microfossils in pieces of sedimentary rock, such as chalk or limestone.

This method shows how you can break down these rocks into particles by soaking them for several days in vinegar.

1. Put some small rock pieces into a jar of vinegar. Leave them until they break down into a sludge.

2. Swirl the mixture around and carefully pour off the muddy liquid, leaving any solid pieces in the jar.

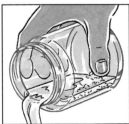

3. Add fresh water to the jar and repeat step 2. Do this several times until the liquid you pour off is clear.

4. Place any solid fragments that remain on a slide. If you are lucky, you will see some microfossils.

Go to **www.usborne-quicklinks.com** *for links to the following websites:*
Website 1 *Advice on collecting foraminifera and some amazing pictures to look at.*
Website 2 *Explore the possibilities of life on Mars and view a Martian microbe.*

*You can find out more about polarized light on pages 69 and 93.

This microfossil has been magnified to 135 times its real size.

Fossil fuels

Oil and gas are called fossil fuels because they are made from the remains of microfossils like the ones on these two pages. The shells of these sea creatures became part of the rock, but their fleshy parts turned into oil and gas. This seeped through the rock to collect in large quantities.

Drilling for oil

Geologists working for oil companies use microscopes to look for microfossils. They cut rock samples into very thin sections so the light can shine through them. If they can see any microfossils, this means that oil may be nearby, and it is worth exploring this area more thoroughly.

Fossilized bacteria

As well as the fossils of tiny sea creatures, geologists have also found examples of bacteria that have turned into rock.

First life

Some of these are among the earliest evidence for life on Earth. The oldest bacteria discovered so far are about 3,500 million years old.

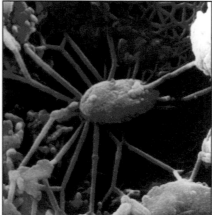

Fossilized bacterium (middle) and sea creatures in a piece of rock.

The pictures above and below show two examples of fossilized bacteria. They were both found in rock samples drilled from deep below the ocean bed.

This fossilized bacterium is about 10 million years old. It is shown here magnified 12,500 times.

Life on Mars?

A rock that had fallen from the planet Mars to Earth was discovered in Antarctica in 1984. Electron microscopes revealed many tube-like structures (shown below in pink) on the rock's surface.

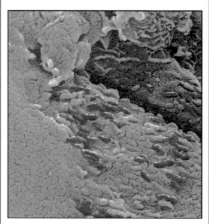

These bacteria-like structures are over 100 times thinner than a human hair.

Some scientists think that these are microfossils of organisms which may have lived on Mars. Others think that bacteria from the Earth contaminated the rock after it landed.

CRYSTALS

Many natural things, from salt to diamonds, are made up of shapes called crystals. When you look at crystals with a microscope, you can see that most have flat sides. You might also be able to see spectacular colours or patterns in them.

Crystal shapes

The cube-shaped crystals in the electron microscope picture below are salt crystals. Like most crystals, they have regular, flat-sided shapes.

The number of sides a crystal has, and its shape, make each type of crystal different. For example, all of the crystal types pictured on these pages have different shapes.

Looking at table salt

If you look at crystals of table salt under your microscope, you will be able to see their square, flat sides. But you may also notice that some grains have rounded corners.

This rounding happens in salt factories where white table salt is produced. The salt passes through several large vats. As the crystals bump into each other in the vats, their sharp edges and corners are often worn smooth.

Crystals of salt, magnified 840 times by an electron microscope.

The crystals above are diamonds. The top ones have been cut by a jeweller. The one in the middle is in its natural form.

Diamonds for ever

Diamonds are found in a rock called kimberlite. They are the hardest minerals known, and are used in industry to make tools, such as the dentist's drill below.

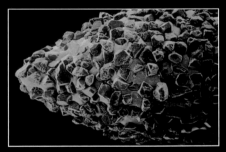

This diamond-studded dentist's drill will cut cleanly through the surface of teeth. It is shown here magnified 75 times.

Experts use microscopes to help them to value diamonds. The most expensive diamonds, like the two at the top of the page, are colourless and have no faults, such as other minerals, in them. They are cut to make sparkling gems for jewellery.

Rainbow crystals

The caster sugar crystals below have been viewed through a polarizing microscope, which uses a particular type of light, called polarized light. As the light shines through these crystals, it splits into different colours, like a rainbow, making the crystals easier to see.

Colourless crystals

Many crystals are difficult to see through an ordinary microscope because they have no colour. You can see them more easily with a polarizing microscope.

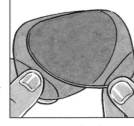

Sugar crystals viewed with polarized light.

Rock crystals

Crystals in a rock viewed with a polarizing microscope, magnified 15 times. To view a rock in this way, scientists cut a section that is thin enough for the light to shine through.

Rocks are made up of crystals of minerals. Above you can see how the various crystals in a rock sample show up in different colours under a polarizing microscope.

These colours help scientists to identify rocks. They can also help them to identify any valuable minerals, such as gold, or fuels that may be in them.

Polarized light

You can view crystal samples under your microscope with polarized light. You will need a cheap pair of clip-on plastic polaroid sunglasses (grey lenses are best). Try looking at grains of sugar – you should be able to see some colourful results.

1. Remove the lenses from the glasses and place one lens on top of the other as shown above.

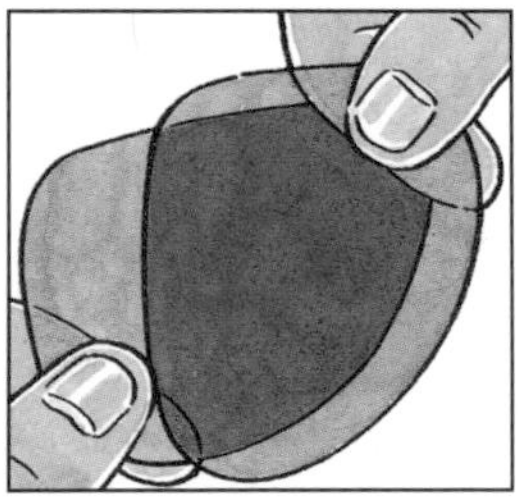

2. Turn the top lens through 90^0 (one quarter turn) so the lenses go dark. Keep them at this angle.

3. Use adhesive tape to attach one lens under the microscope stage above the light or mirror.

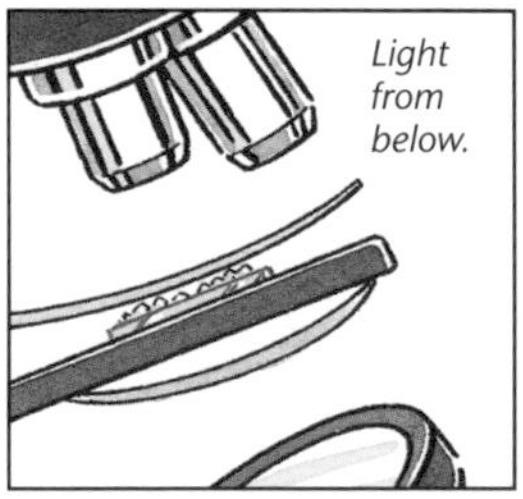

4. Place the other lens on top of the crystals. Make sure the angle of the lenses is the same as in step 2.

Light from below.

Go to www.usborne-quicklinks.com for a link to a website where you can discover what different types of sugar look like under the microscope and find some sugar facts too.

MORE CRYSTALS

You can easily see the crystal shapes in salt and sugar. But the substances which are shown on these pages have crystals that are more difficult to see, until you view them with polarized light*.

Kaleidoscope patterns

The stunning patterns on the right are crystals of the pain-relieving drug aspirin, under a polarizing microscope. The aspirin was dissolved in a liquid and allowed to dry, leaving crystals behind. All the crystals shown on these pages were made by this method.

Different crystals make different patterns under polarized light. This can help scientists to identify the substances they are examining.

Paracetamol crystals

On the left, for example, you can see some crystals of paracetamol, another pain-relieving medicine. Its pattern is very different from the aspirin crystals. This is mainly because it is made from a different chemical.

Salicin crystals

These are crystals of a type of sugar called salicin, which is found in the bark of willow trees. The pattern of the salicin crystals is similar to the pattern of the aspirin crystals. This is because aspirin contains a chemical made from willow tree salicin.

Aspirin crystals, magnified 50 times with a polarizing microscope.

*See page 69 for more about polarized light.

Crystal solutions

Most of the crystals shown on these pages were made from medicines, or chemicals that are difficult to obtain. But you can make your own crystals from harmless substances that are easy to get hold of.

Below, for example, you can see some crystals of glucose sugar viewed with polarized light. They were made using the method explained at the bottom of the page.

Crystals of glucose can produce some interesting effects under polarized light. These ones are magnified 190 times.

Art and science

These are crystals of a cancer drug called taxol. They have been viewed by a polarizing microscope and magnified about 60 times. As well as showing the crystal structure of the drug, this microscopic view almost looks like a work of art.

Many varied and spectacular effects can be achieved with polarized light and crystals. Some enthusiasts experiment with different ingredients and methods of making crystals, to create individual works of art.

It is possible, for example, to vary the size of crystals you make. You can obtain smaller crystals by drying the sample quickly on a radiator. If you dry the sample more slowly, larger crystals will form.

Making crystals

You can use this method to make crystals of glucose or dextrose to look at with polarized light (see page 69). You will need a glucose or dextrose tablet, which you can buy in packets from a pharmacist or health food shop.

1. Place one teaspoon of warm water in a cup. Crush one tablet and stir it in, until it has dissolved.

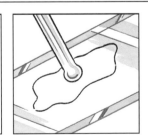

2. Use a glass rod to spread a few drops of the glucose or dextrose solution onto a microscope slide.

3. Leave the solution to dry, either at room temperature or somewhere warmer.

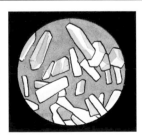

4. Look at the crystals using polarized light. If you are lucky, you may see beautiful colours in them.

Go to **www.usborne-quicklinks.com** for links to the following websites:
Website 1 *See incredible pictures of crystals and find out how they were achieved.*
Website 2 *Take an unusual journey through a land made of crystal images.*

TESTING METALS

Although metal surfaces can look smooth and shiny, microscopes reveal that they are made up of minute, rough-edged crystals. In industry, metal scientists use microscopes to check metals for cracks and holes, which may make the metal break and cause an accident.

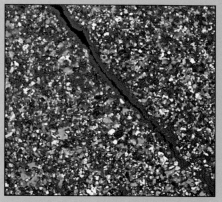

Optical microscope image revealing faults in a piece of copper, shown over 240 times its real size.

Identifying metals

The picture below shows a cross-section of two kinds of metal that have been joined together.

At the bottom is a piece of iron. It is made up of tiny, tightly-packed crystals. The edges of the crystals are straight and easy to see.

The metal at the top is steel. This is made by mixing iron with a substance called carbon, to make it stronger. Metals which are mixed together are called alloys. You can see here how the crystal structure of steel is not as regular as the crystal structure of iron.

Optical microscope image of iron and steel, magnified over 140 times. The colours you can see are dyes that were added to show the details more clearly.

Tube testing

Cracks appeared in this alloy during tests to see how much pressure it could withstand. (Shown magnified 200 times.)

When new machines are designed, engineers try out different metals to see which one works best. The alloy shown above was used to make a tube in a piece of factory machinery.

The tube was made from three different metals and designed to carry hot liquids at high pressure. When it was tested, the tube cracked. Microscopes revealed that some of the crystals had torn apart. This meant that it was not safe to make the tube from that particular alloy.

Black beads

Scientists sometimes need to investigate metals to find out the cause of accidents. The black beads you can see above are actually tiny holes which formed along the edge of the crystals in a piece of copper, when it was heated.

Big bang

The copper was used to make parts in electrical fuses which were put into a power station. The holes in the copper seriously weakened the fuses and led to an explosion.

Joining metals

Engineers use microscopes to test different ways of joining metals together. For example, on the right you can see a piece of steel that has been joined to a piece of brass, using a method called explosive welding.

Making waves

One piece of metal was fired into the other. It hit with such force that the metals acted like liquids, and made the waves you can see here. The join was examined with a microscope to make sure that there were no cracks which could cause the metals to come apart.

Optical microscope image of a join between steel (shown in grey-yellow) and brass (brown), magnified 40 times.

Diamond dents

The optical microscope picture below shows a series of dents in a metal surface.

A piece of metal, dented in tests to find out its hardness.

The dents were made by a special microscope, which engineers use to measure the hardness of a metal. In these microscopes, a pyramid-shaped diamond is fixed in front of the objective lens.

A weight then drives the diamond into the metal, making a dent. By measuring the depth of this tiny dent, engineers can find out how hard the metal is.

MICROMACHINES

The ability to make machines smaller and smaller is one of the most exciting areas of science. Computers already use minute machines called chips, but scientists think that other types of micromachines will also become commonplace over the next ten years.

This micromachine is shown 600 times larger than life. It operates a car air bag.

Chips

The pictures below and right are enlarged views of the tiny electronic circuits in a chip. Chips are found in every computer. They control the way the computer operates, and also store information. Wires around the edge of the chip connect it to other chips in the computer.

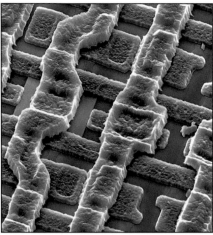

The surface of a chip, magnified 3,100 times. The orange specks are dust.

The chips are made very small so that lots of information can be packed into the computer. Their small size also allows a computer to work faster. This is because electric messages passed between the chips do not have as far to travel.

Computer chips are currently the most common kind of micromachine.

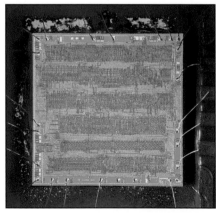

A chip, magnified about seven times.

These computer chips are made from a hard but flexible substance called silicon.

Accident detector

Above is another type of micromachine called an accelerometer. This one is only three millimetres wide. It operates the air bag in a car.

Vibrating teeth

In an accident, a car might speed up or slow down very rapidly. When this happens, the interlocking teeth you can see in the picture shake rapidly, and generate a small electrical signal. This makes the air bag fill with air to protect the driver.

The air bag is hidden in the steering wheel.

During an accident, it fills up with air in an instant.

*Go to **www.usborne-quicklinks.com** for links to the following websites:*
Website 1 *Watch movies, see pictures and read information about micromachines, imperceptible to the human eye.*
Website 2 *Follow an interactive tutorial to find out how a microprocessor, or computer chip, is made.*

Micro engines

In the electron microscope picture below you can see a minute electric motor which has been engraved on a wafer of silicon.

The tiny gear (shown in yellow) moves the larger gear at a speed of one complete turn a second.

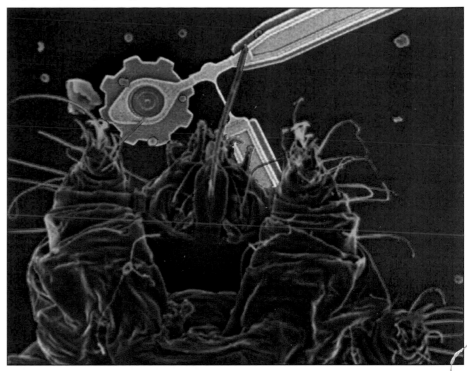

This electron microscope picture shows a dust mite crawling over the motor shown on the left. At its real size, the mite is almost invisible to the human eye.

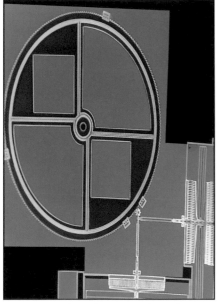

A micro engine, shown 28 times its actual size. The tiny gears are smaller than the width of a human hair.

At the moment, machines such as this are still being developed. One possible use for this kind of technology is to operate a tiny pump within the body. This would release a dose of a particular drug into the bloodstream at regular intervals.

Compact car

This working model of a car was built by Toyota in 1997, to celebrate the company's sixtieth anniversary. It is less than five millimetres long. Although the model was built for fun, it shows how it is possible to build extremely accurate, intricate machines at minute sizes.

Toyota's mini-car takes almost a day to travel 1.6km. It has its own engine, but is powered by a battery which is too big to fit inside the car.

NANOTECHNOLOGY

The tip of a scanning tunnelling microscope, magnified 2,300 times.

Scientists can now pick up individual atoms and molecules (groups of atoms) and move them around. This means that in the future it may be possible to build complex machines so small that they are invisible to the naked eye. This area of science is called nanotechnology.

Tiny tip

Atoms can be picked up with the tip of a type of electron microscope called a scanning tunnelling microscope. You can see one of these tips at the top of the page.

The microscope tip can only move the atoms in a vacuum, when there is no air or any other gas in the space around them.

Moving atoms

The series of pictures below shows a cluster of molecules being moved with a scanning tunnelling microscope tip to make a new shape. The circle the atoms form in the final picture would never normally be found in nature.

Small world

Atoms are the basic building blocks of any substance. They are so small that about 15 million of them would fit onto a pinhead.

Atomic machines

Scientists hope that by learning to manipulate atoms, they will be able to build more complex structures.

At the moment, the wheel of atoms on the left only exists as a computer drawing. Each ball represents a single atom. One day, components like this could be built for real and used to make miniature machines, such as engines, pumps and robot arms.

A computer-generated image of a wheel made from individual atoms.

The molecules at the bottom of these pictures are being moved by a scanning tunnelling microscope tip to form a circle.

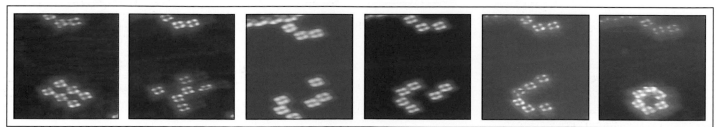

Atom art

The images on this page are all taken with an electron microscope. They show the kind of arrangements that researchers have been able to make with atoms.

Scientists at the electronics company IBM produced this image of their company's logo using atoms of xenon gas.

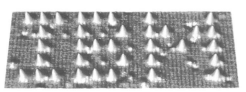

Each blob is a single atom.

The picture below shows the Japanese word for atom. This was also produced by IBM scientists, using iron atoms on a copper background.

These atoms spell "atom" in Japanese.

Atom abacus

The false-colour electron microscope picture below right shows an abacus which uses individual molecules as counters.

Carbon balls

Each counter is made from 60 carbon atoms, which are placed in grooves on a copper surface.

The counters can be moved along the grooves with the tip of a scanning tunnelling microscope.

Abacus made with molecules, showing the numbers zero to ten. Each counter is less than a nanometre (one millionth of a millimetre) across.

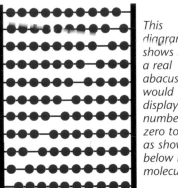

This diagram shows how a real abacus would display the numbers zero to ten, as shown below in molecules.

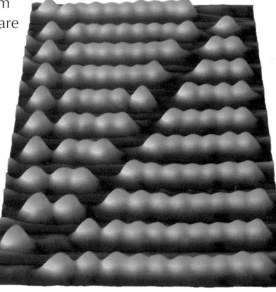

Crystal guitar

This model of a guitar is made from silicon crystal, the same material used to make microchips. It is about the size of a single human body cell. It even has six strings. Each one is one thousandth the width of a hair.

Why do they bother?

The reason why scientists make devices such as this is to investigate the problems of building such minute and intricate machines.

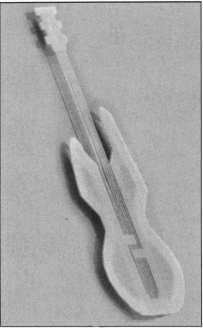

This guitar was made at Cornell University, New York, in 1996.

Go to **www.usborne-quicklinks.com** for links to the following websites:
Website 1 Listen to nano-sized instruments and see how nanobots may be used to fight disease.
Website 2 Try building a virtual atom online and learn more about what real atoms are made of.

A NANO FUTURE?

At the moment, research into nanotechnology is directed toward projects which have a very good chance of actually happening.

Scientists are currently working on producing even smaller computer chips than those that are used today.

Other areas of research include building lightweight spacecraft (which would need less fuel to launch them up into space), hearing aids that will fit unnoticed inside the ear, and tiny components for military devices such as mini spy planes.

More to come?

Nanotechnology's most enthusiastic supporters foresee far more exciting possibilities for this new science. They predict the invention of tiny, complex nanomachines, such as the miniature device you can see in the illustration below.

Copying nature

Some scientists think that if they can learn to understand how nature works at the most microscopic level, they should then be able to copy it. This could lead to nanomachines

taking basic ingredients like air, water and soil and using these to reproduce themselves, in much the same way as plants do.

Not so clever

Most scientists are more doubtful, and think it will be impossible to produce such minute machines. They compare the more extreme predictions for nanotechnology with the fantastical dreams of some scientists in the 1960s. These scientists imagined that it would soon be possible to colonize distant planets, or to invent computers that could think like humans.

Out of control

Worst of all, say the doubters, machines that can reproduce themselves could go out of control. This could lead to a very frightening situation where nanomachines reproduce like a cancer – engulfing everything in their path.

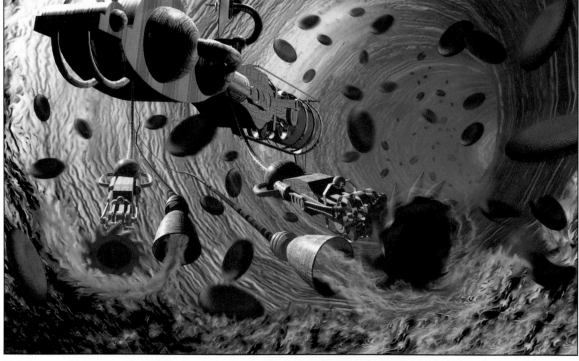

This is an artist's impression of the type of nanomachine that could be built in the future. A minute submarine-like device is trawling a human artery and sucking up dangerous blood clots (shown in green).

Go to **www.usborne-quicklinks.com** for a link to a website where you can look at a gallery of fascinating pictures by artists that show how nanorobots may one day be used in medicine.

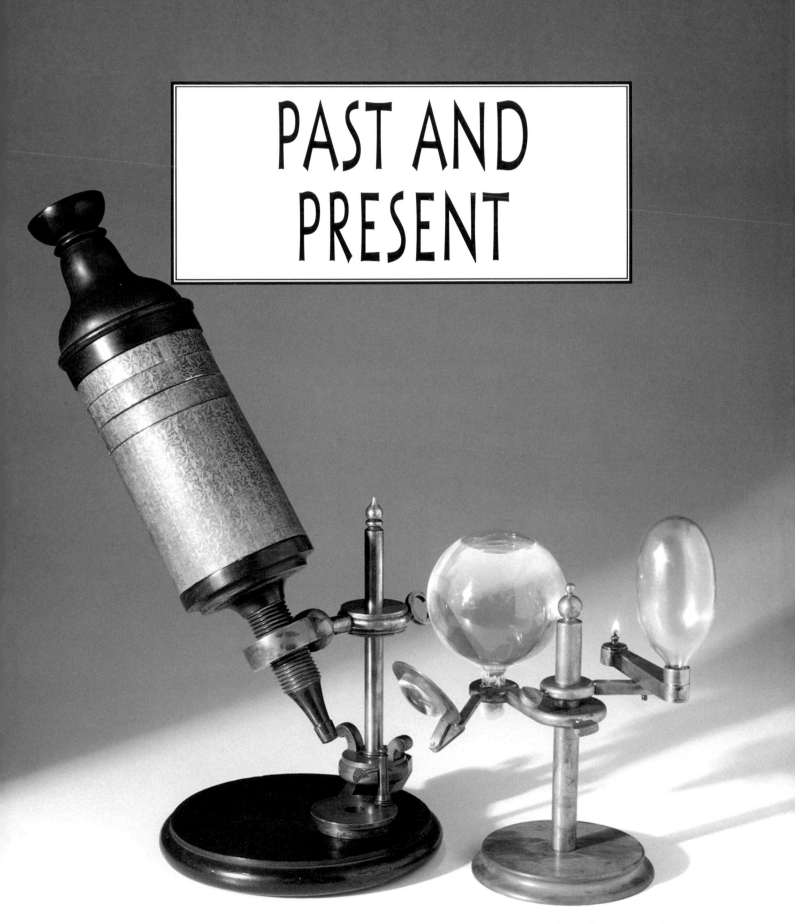

PAST AND PRESENT

Copy of a seventeenth-century microscope.

EARLY MICROSCOPES

Scientists began making microscopes in the early seventeenth century. No single person can claim to have invented the instrument, as several types came into use around the same time. These devices were crude, but they enabled some important discoveries to be made.

Eyepiece

One of the first microscopes. This instrument was owned by Robert Hooke.

Objective lens

Specimen placed here

Hooke's book

The picture below of the stings on a nettle leaf was drawn by an English scientist named Robert Hooke. The illustration appeared in a book called *Micrographia* which was published in 1665. It contained drawings of specimens he had looked at with his microscope. Subjects included body lice, plant cells, fungi and snowflakes.

Hooke's drawing below shows the structure of cork. His microscope enabled him to discover cells – the minute building blocks that make up plants and animals.

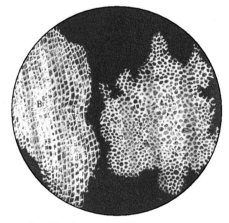

Cork cells, drawn by Robert Hooke. He was the first scientist to use the word "cell" to describe the units which living things are made out of.

Hooke's drawing of a flea is almost as accurate as the electron microscope view on page 52.

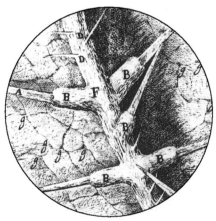

An engraving of Hooke's drawing of stinging hairs on a nettle.

Hooke's microscope enabled him to discover how nettles sting. He was able to observe poison travelling from hollow stinging hairs into his finger.

Hooke's microscope

Hooke made the drawings on this page while looking through a microscope like the one above. It used three glass lenses to magnify the objects. These were lit with an oil-fed flame. The light was concentrated into a bright spot by a water-filled ball. (You can see a picture of this on page 79.)

Go to **www.usborne-quicklinks.com** for links to the following websites.
Website 1 *Browse a brief illustrated history of the optical microscope and the innovations of key scientists.*
Website 2 *A good overview about the history of magnification beginning with the ancient Egyptians.*

Waterscope

You can make a microscope which will give a similar view to early ones. They used glass lenses, but you can use a drop of water as a lens. You will need a pin, a pencil, a small piece of cardboard, some clear plastic film and some modelling clay.

1. Put the cardboard on some modelling clay. Make a hole in it using a pin, then a pencil, as shown.

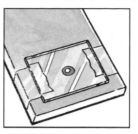

2. Remove the clay. Place a piece of clear plastic film over the hole and stick it down with tape.

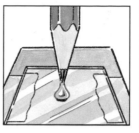

3. Dip a pencil in a glass of water, then use it to place a small drop of water over the hole.

4. Hold your water microscope close to this page. The print should appear greatly magnified.

Blurred images

Lenses in early microscopes were of much poorer quality glass than the lenses we use today. They produced images which were often distorted, and had a ring of blurred colour around the edge.

Surprisingly accurate

Despite the problems, the pictures on these two pages show that many of these early observations were surprisingly accurate. But there were exceptions. Compare this drawing of a sperm cell on the right with the electron microscope picture of one on page 31.

This picture was produced by a Dutch microscope maker named Nicolaas Hartsoeker. He was convinced that sperm cells contained little babies.

Hartsoeker's drawing of a sperm cell.

Primitive equipment

One of Leeuwenhoek's microscopes.

This screw enables the viewer to bring the specimen into focus.

The specimen to be examined is placed on this spike.

There is a single lens here. Leeuwenhoek looked at the specimen from the other side of this hole.

This microscope was made in 1686 by a Dutchman named Antoni van Leeuwenhoek.

Using this primitive equipment, Leeuwenhoek discovered many previously unseen things, such as red blood cells, and bacteria which he noticed in the sticky white coating he scraped off his teeth.

This is the spike on the microscope where the specimen is attached.

Leeuwenhoek observed this fly with his microscope.

Go to **www.usborne-quicklinks.com** for links to the following websites:
Website 1 See images of microscopes through the ages, from 16th century examples to the latest models of today.
Website 2 Visit an online exhibition featuring a collection of over 40 historical microscopes.

THE WAR ON GERMS

Louis Pasteur used a microscope to prove there was a link between germs and disease.

Lenses in early microscopes were poor quality and only magnified objects around 200 times. Improvements in the early nineteenth century increased magnification to 2,000 times. The invention of electron microscopes in the 1930s made it possible to magnify up to a million times.

Breakthrough

Better microscopes were developed in the early nineteenth century by men such as Joseph Lister. You can see him pictured below with one of his instruments.

In 1830, Lister made lenses which produced much clearer images of the objects being studied. His lenses also made it possible to view objects at a much higher magnification than earlier microscopes.

In this 1827 cartoon, a horrified London lady peers through a microscope at a drop of germ-infected drinking water from the River Thames.

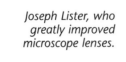

Joseph Lister, who greatly improved microscope lenses.

Deadly connection

By the early nineteenth century, many people were beginning to realize that there was a definite connection between germs and disease. The cartoon above shows that people were also aware that diseases could be carried in infected water.

Joseph Lister's improved microscopes finally enabled scientists to show this was true.

Louis Pasteur

Seen below peering through a microscope in his laboratory, is nineteenth-century French scientist Louis Pasteur. He used a microscope to prove beyond doubt what most scientists had long suspected – that germs did carry disease from one living creature to another.

Sour milk

Pasteur's microscope also enabled him to discover why milk, beer and wine go sour. He saw how tiny bacteria in these liquids multiplied in great numbers, causing them to "go off".

Pasteur used a microscope to discover why beer, wine and milk go sour.

Pasteurization

Pasteur also discovered that by gently heating these liquids, it was possible to destroy many of the bacteria in them, and increase the time they stay fresh. This process is called pasteurization.

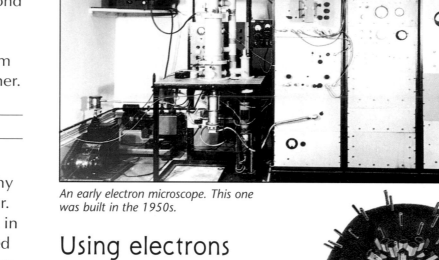

An early electron microscope. This one was built in the 1950s.

Using electrons

In the 1930s, scientists invented the electron microscope. Ordinary optical microscopes use light to study objects. Electron microscopes use tiny particles of atoms called electrons. (See page 9 for an explanation of this.)

Smallest of all

Electron microscopes opened a whole new world to scientists. These microscopes made it possible to study some of the smallest objects on Earth, such as viruses and atoms. On the right you can see some of the first illustrations of viruses, as seen by an electron microscope in the 1950s. These viruses are impossible to see with an optical microscope.

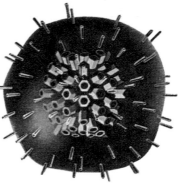

The virus above causes cold sores.

This virus attacks bacteria.

This virus causes a disease called smallpox.

Go to **www.usborne-quicklinks.com** for links to the following websites:
Website 1 Read more about the lives and innovations of Joseph Lister and Louis Pasteur.
Website 2 Try a fun quiz about good and bad bacteria, with close-up images.

MICROSCOPES TODAY

Our ability to magnify things has taken a huge leap forward in the twentieth century.

In the 1900s, scientists used one basic type of microscope – the optical microscope. Although these instruments were greatly improved models, they worked in much the same way as the very first microscopes of the 1590s.

Spoiled for choice

Optical microscopes are still very useful, but scientists today have over 40 different other types of microscopes to choose from. These work in very different ways.

Atom images

One of these microscopes – the scanning tunnelling microscope – is so powerful that it can produce an image of a single atom.

Controlled by a computer, this microscope brings a very sharp tip extremely close to the surface it is studying. Then it records minute changes in electrical forces between the tip and the surface, and uses these measurements to produce an outline picture.

(You can see a picture of this kind of tip, and individual atoms, on pages 76-77.)

Sound and vision

Another type of microscope, called an acoustic microscope, uses sound waves to "see" through solid materials like metals. It can focus on a tiny area a fraction of a millimetre across.

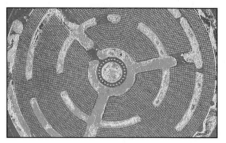

Sound waves produced this picture of the top of a screw.

The picture above is produced by an acoustic microscope. It shows a close-up view of a screw in a nuclear reactor.

Other modern microscopes use X-rays, magnetic forces or differences in temperature to build a magnified picture of the object they are examining.

Computer screens

Most modern microscopes show the object they are studying on a computer screen. This image can be altered to make it clearer. Many of the pictures in this book, for example, are computer screen images which have had colour added to make them easier to understand.

A scanning tunnelling microscope. The object being studied is placed in the tall chamber on the right, and can be seen on the computer monitors in the background.

Go to www.usborne-quicklinks.com for a link to a website where you can try an online simulation of a scanning tunnelling microscope.

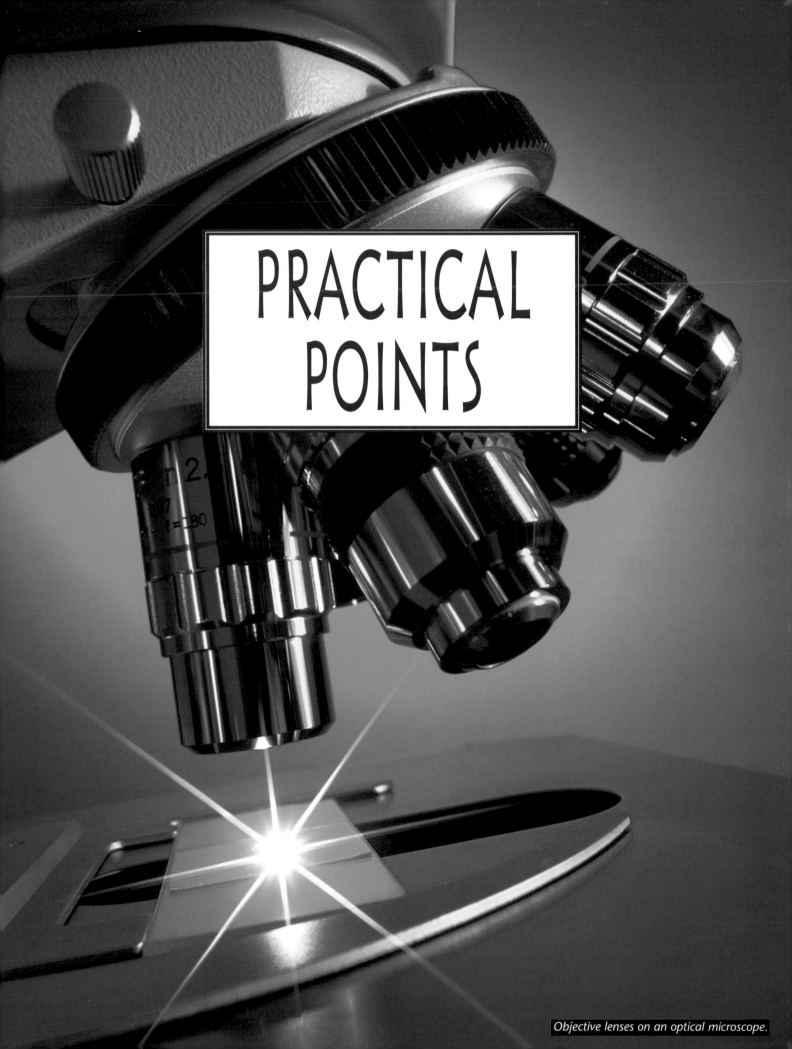

PRACTICAL POINTS

Objective lenses on an optical microscope.

BUYING A MICROSCOPE

I f you enjoy exploring the microscopic world, you may decide that you want to own a microscope. All of the ones shown here are optical microscopes, and work in a similar way to the one shown on page 8. There are many different models to choose from.

A simple microscope

What to buy

If you are a beginner, you don't need to buy an expensive microscope. It's better to buy a good quality basic model. This can be upgraded at a later date with accessories such as different eyepieces, or higher power objective lenses.

Best results

A good quality, low-power microscope will give you much better results than a cheap microscope that claims to magnify up to 600 or even 1,000 times. The quality of the lenses and the mechanism inside most cheap models means that the images you see are likely to be blurred and disappointing.

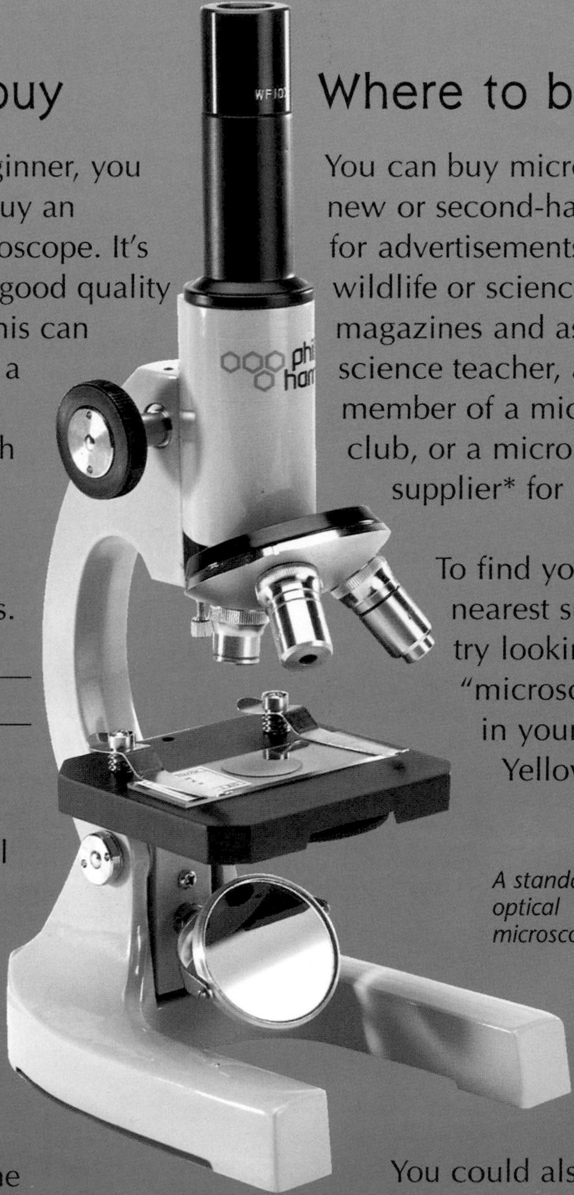

Where to buy

You can buy microscopes new or second-hand. Look for advertisements in wildlife or science magazines and ask a science teacher, a member of a microscope club, or a microscope supplier* for advice.

To find your nearest supplier, try looking up "microscope" in your local Yellow Pages®.

A standard optical microscope

You could also try searching under "microscope+accessories" on the Internet. If at all possible, try out a microscope before buying it.

Good quality

The model above is called a simple microscope. It is inexpensive and produces a good quality image at low power. It comes with a small light that can be used to light objects from above or below. This is a good microscope for a beginner as it is easy to use.

Different magnifications

Alternatively you might want to buy a standard optical microscope such as the one shown on the left. It has three objective lenses, so it can be used to view objects at different magnifications.

Light from below

At the base of the microscope is a mirror which can be tilted to reflect light from a lamp onto the object you are looking at. You can also buy microscopes with built-in lamps to light objects from below, but these are likely to be more expensive.

*See list of addresses on page 95.

Seeing in stereo

This stereomicroscope below is really a pair of microscopes, with one eyepiece for each eye.

It enables you to see the object you are looking at in three dimensions, because each eye has a slightly different view.

Although the magnification is quite low (about x10 to x35), the images that you see can be spectacular. This type is ideal for looking at whole objects, such as insects, leaves, flowers, fabric, feathers, coins and sandpaper.

Viewed with a stereomicroscope, this moss sample appears in three dimensions.

Out and about

Pocket microscopes like this one on the right are light and small, so that they can be carried around outdoors. This model can magnify objects either 80 or 200 times. It contains a small light to illuminate specimens.

Accessories

Some optical microscopes have several eyepieces and objective lenses to choose from, each with a different magnification. Make sure that you buy the right parts for your particular microscope.

x40   x10   x4   x100

The objective lenses above can all be attached to the same microscope.

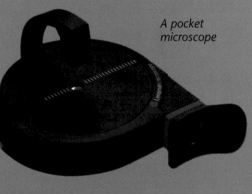

A pocket microscope

Taking care of your microscope

You will need to keep your microscope in good condition, if you want to get the best results from it.

Dust and greasy marks on the lenses will make the images you see look less clear. You also need to be careful when you clean a lens because dust particles can easily scratch its surface. You can protect your microscope from dirt in the following ways.

1. Remove dust and grease from the lenses by wiping them gently with a lens tissue or a soft brush. You can buy lens tissues from an optician or a camera shop. Many camera shops also sell blower brushes, which puff air to blow away the dust.

You squeeze the rubber bulb to blow air onto the lens.

Blower brush

2. Always leave an eyepiece in the top of the body tube. This stops dust from settling inside the tube and on the objective lenses.

Eyepiece

3. Always keep a dust cover over your microscope when you are not using it.

Dust cover

Go to **www.usborne-quicklinks.com** *for links to the following websites:*
Website 1 *Pick up tips on how to choose the right microscope as a first time buyer.*
Website 2 *Browse an online guide to find out what you should know before you decide to buy a microscope.*

87

EQUIPMENT

Here is a list of equipment that you will find helpful when using your microscope. Some items can be bought from a pharmacist or hardware store, and the the rest should be available from a microscope supplier (see page 95).

① Measuring jug

② Glass jar with holes in the lid, for storing pond and sea water. The lid holes let in air so that living things in the water can stay alive.

③ Different sized jars with lids, for storing objects you want to look at.

④ Lens tissues and a blower brush for cleaning your microscope lenses.

⑤ Cover slips for placing over specimens on slides to stop them from drying out. Be careful, as these are very thin and can be broken easily.

⑥ Ring and ⑦ cavity slides for examining life in a drop of water. A ring slide is an ordinary slide which has a glass ring glued to it.

You can buy glass rings from some microscope suppliers. Alternatively, you could draw a circle of glue on a plain slide to keep the water in place.

⑧ Plain slide for looking at most types of samples.

⑨ Stage micrometer and ⑩ eyepiece graticule to help you measure your specimens.

⑪ Stains (see page 91).

⑫ Watch glass or see-through plastic container for preparing specimens.

⑬ Plastic petri dish, which can be thrown away after use.

⑭ Polarspex. An inexpensive and effective way of creating polarized light. Two small sheets of polarizing material are held in stiff cardboard. Simply slide your crystal samples between the sheets to see vivid polarized colours.

⑮ Scalpel and sharp kitchen knife, and a plastic cutting board. (Always be very careful with sharp blades.)

⑯ Mounting needles for arranging insects on slides.

⑰ Tweezers. It is useful to have a pair with blunt ends and a pair with pointed ends.

⑱ Dropper and glass rod, for placing liquids onto slides or petri dishes.

⑲ Notepad or other paper.

⑳ Cotton buds

㉑ Sticky labels for recording details on slides or petri dishes.

㉒ Disposable plastic gloves

㉓ Absorbent paper, such as blotting paper or sheets of paper towel.

㉔ Sticky tape, scissors, a fine, soft paintbrush, pencils and pens.

㉕ Disinfectant for cleaning slides and other equipment.

㉖ Adjustable lamp for lighting specimens.

ADVANCED TECHNIQUES

Most of the microscope experiments in this book are fairly easy to do. These two pages tell you about some more complex techniques. For these you will need extra equipment, which you can buy from a microscope supplier (see page 95).

Cutting sections

The best way to cut slices, called sections, from an object you want to look at, is with a tool called a micro-tome. The simplest kind, a hand microtome, is ideal for cutting the soft parts of plants, such as stems or leaves.

A hand microtome

The specimen is placed in this hole.

The screw pushes the specimen up through the hole.

You will need: a hand microtome and blade; a 2cm length of plant stem, such as a tulip stem; a carrot; a dish of water; and a small paintbrush.

> WARNING:
> Microtome blades are very sharp. Always be extremely careful when cutting sections.

Using a microtome

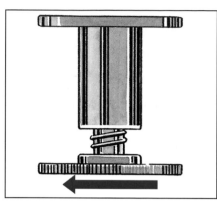

1. Hold your microtome in one hand and turn the screw to wind it down to the bottom of the hole.

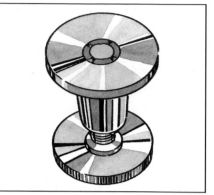

2. Put the specimen in the top of the hole. If it is smaller than the hole, use strips of carrot to wedge it in.

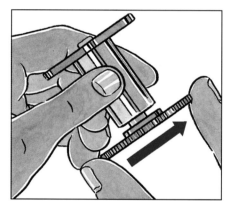

3. Gently turn the microtome screw the other way until it starts to push the specimen out of the hole.

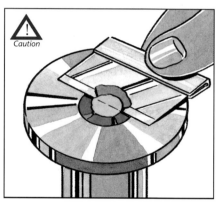

Caution

4. Carefully cut some sections. The thickness of the section will depend on how far you turn the screw.

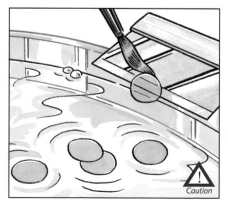

Caution

5. Use the paintbrush to slide each section off the blade into the dish of water. This helps to keep them fresh until you are ready to look at them.

Coloured light

Viewing your sample with coloured light can make it easier to see. You can buy coloured cellophane sheets from a stationer to do this.

Hold the cellophane below the objective lens or cut it to fit your mirror. Mixing the sheets gives different colours.

Choosing stains

To show more detail in your cell specimens, you can add different types of stains. (Two methods of staining are shown on pages 41 and 43.) Food colouring makes the sample look more solid. Other stains only colour certain parts of cells. Here are two stains that are easy to get hold of.

Iodine can be bought as "tincture of iodine" from a pharmacist. It turns the nucleus brown and cytoplasm yellow. Iodine also stains the parts of cells which store food in the form of sugars, turning them dark blue in plant cells, and red in animal cells.

Cochineal is a food colouring you can buy from a supermarket. It contains a pink stain called carmine, which can turn the nucleus slightly darker than cytoplasm.

Gelatin mounts

If you want to make specimens last longer than they do in temporary (wet) mounts, you can use the following method. This type of mount is called a semi-permanent mount.

You will need gelatin and glycerin, which you can buy from a supermarket or pharmacist. You will also need some gum arabic and enamel paint, which you can buy from a model or art shop.

Go to www.usborne-quicklinks.com for a link to a website where you can find out more about using a microtome.

Making a semi-permanent mount

1. Add a small sachet of gelatin to five teaspoons of water and five teaspoons of glycerin in a jar.

2. Sit the jar in a larger bowl of hot water and wait until the gelatin dissolves in the water and glycerin.

3. Put a specimen on a plain slide. Use a dropper to place four drops of the warm gelatin mixture onto it.

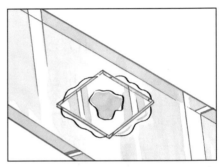

4. Put a cover slip over the top and leave the gelatin to set. Try not to trap air bubbles under the cover slip.

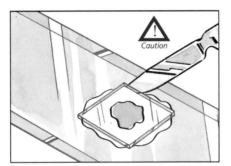

5. With a scalpel or sharp knife, cut away any gelatin which oozed out from under the cover slip.

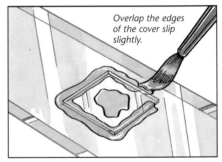

Overlap the edges of the cover slip slightly.

6. Use a small paintbrush to put a layer of gum arabic around the edge of the cover slip. Leave it to dry.

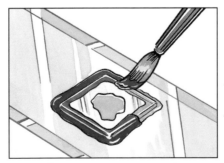

7. Brush two coats of enamel paint over the gum arabic, overlapping the edges slightly as shown. Allow the paint to dry between coats.

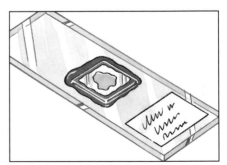

8. Label your mounts, giving the date and information about the specimen, such as where it was collected and how it was stained.

GLOSSARY

This glossary explains some of the important words you will come across when reading about the microscopic world. Words in *italic type* have their own entry elsewhere in the glossary.

alloy A mixture of two or more metals, or of metals and non-metals.

antennae (singular **antenna**) Feelers on the head of an insect that detect touch, smell or vibrations.

antibiotic A type of medicine that destroys *bacteria*.

antibodies Chemicals made by a type of white blood *cell* which fight off *germs* that cause disease.

atom The smallest part of any substance, such as gold, that is still recognizable as being made of that substance. Everything in the world is made up of atoms.

bacteria (singular **bacterium**) Microscopic living *organisms* that live all around you and inside you. Many are harmless or useful, but some can cause disease.

binocular microscope A type of *optical microscope* which lets you see the same image with both eyes, making viewing more comfortable.

body tube The part of an *optical microscope* which contains the *eyepiece* and *objective lenses*.

cancer A disease caused by faulty body *cells*, which continue to grow and divide after they should have stopped. This forms a lump of cells, called a tumour, which can destroy the healthy cells around it.

cell The smallest unit in a living plant or animal that can work on its own. (For example, a cell can make its own food and reproduce without help from other cells.)

chromatic aberration A problem with some *lenses*, where a blurred rainbow effect appears around the edges of the image.

chromosomes Microscopic bundles that appear in a *nucleus* during *cell* division. These bundles contain minute units called *genes*.

coarse focusing control A knob on an *optical microscope* which brings the *object* into *focus*. (See also *fine focusing control*.)

colony A group of the same type of *bacteria*.

crystal A solid, regular shape with flat sides and angular corners.

cytoplasm The watery, jelly-like substance inside a plant or animal *cell*, which contains the *organelles*.

dark ground illumination A lighting technique for showing detail in *transparent objects*, such as some water insects. Only light which has reflected off the *object* is allowed into the *objective lens*. This method produces a view of bright *objects* on a dark background.

depth of field/focus The range over which a *magnifying lens* or microscope will produce a *focused* image. The greater the depth of field (for example in a *stereomicroscope*), the more three-dimensional the image seems.

DNA A complex chemical called deoxyribonucleic acid, from which *chromosomes* and *genes* are made.

electron A minute particle of an *atom*, used in *electron microscopes*.

electron microscope A type of microscope that uses *electrons* to make a magnified image of an *object*.

eyepiece The part of an *optical microscope* that you look through. It contains a *magnifying lens*.

eyepiece graticule A piece of clear film or glass with lines marked on it which fits into a microscope's *eyepiece*. Often used with a *stage micrometer*, to measure an *object*.

field of view The whole area that is visible under a microscope.

fine focusing control A knob on an *optical microscope* for changing the *focus* by minute amounts. Found on high-power microscopes. (See also *coarse focusing control*.)

focus The state of an image when it looks sharp and clear. (It is then described as being "in focus".)

fossil fuel A fuel, such as oil or gas, that has formed over millions of years from the remains of ancient creatures or plants.

fungi (singular **fungus**) A type of *organism* that has no leaves, roots or flowers. Moulds, mushrooms and toadstools are examples of fungi.

gene Tiny chemical messages in a *chromosome* which tell a *cell* what to do. They affect body characteristics such as eye and hair colour.

germ A microscopic *organism*, such as a *bacterium*, *fungus* or *virus*, that causes disease.

hyphae (singular **hypha**) Tiny thread-like hairs from which all parts of a *fungus* are made.

lens A piece of curved glass or plastic which is used to bend rays of light, to make an *object* look larger than it really is.

magnetic lens Magnets inside an *electron microscope* which bend streams of *electrons* to form a narrow electron beam.

magnification How much an *object* has been *magnified*. For example, magnification x200 makes an object appear 200 times larger than its real size.

magnify To make something look bigger.

magnifying power The *magnification* that can be achieved by a *lens* or microscope. This can be expressed as x20 or x40, for example.

microbe A microscopic living *organism*, such as a *bacterium*, mould or *virus*.

microchip A tiny piece of silicon that has thousands of electronic circuits on it, used in computers and other electronic equipment. It is sometimes known as a chip.

microfossils The remains of tiny water plants and animals that lived millions of years ago, and have turned to rock.

microsurgery Intricate surgery on delicate parts of the body in which the surgeon uses a microscope and miniature surgical instruments.

microtome A tool used to cut thin sections of a *specimen*.

minerals Natural substances formed inside the Earth. They are made up of simple chemicals called elements, which cannot be broken down into any other substances. Some minerals, such as gold, are made up of just one element. Others, such as quartz, contain two or more.

molecule The smallest *particle* of a substance that can exist on its own. A molecule is usually made of two or more *atoms* that have bonded together.

multi-ocular microscope A microscope that has two or more *eyepieces* so that more than one person can look through it at once.

nanotechnology An area of science that intends to make minute machines from individual *atoms* and *molecules*.

nosepiece The part of an *optical microscope* where the *objective lenses* are attached.

nucleus The round or oval part inside a *cell* which controls everything that happens in the cell.

object The thing you want to look at with your microscope.

objective lens The *lens* on an *optical microscope* which first *magnifies* the *object*.

optical microscope A microscope that uses glass *lenses* to make *objects* look bigger.

organelle Any of the minute parts inside a plant or animal *cell*.

organism A living plant, animal, *fungus*, *bacterium* or *virus*.

particle An extremely small *object* or piece of any material.

photomicrograph A photograph taken of an image made by an *optical* or *electron microscope*.

plankton Microscopic plants and animals that float near the surface of areas of water such as lakes or oceans.

polarized light The light waves in polarized light vibrate only in one direction, for example up and down. This is different from ordinary light waves which vibrate in many directions. A filter made out of polarizing material only lets through light vibrating in one direction, so it turns ordinary light into polarized light. If you look at *crystals* with polarized light, you may see colourful patterns in them.

polarizing microscope A type of microscope which uses *polarized light*. Used mainly for looking at *crystals*, plastics and metals, for example, to find weaknesses in them.

pollen Fine grains of yellow powder found in flowers. Pollen contains the male reproductive *cells* which are needed to make new plants grow.

preferential stains Dyes that change the colour of some, but not all, of the features in a *specimen*.

reflection The way light bounces off something, such as a mirror.

refraction The way light rays bend when they pass from one material to another, for example from air into glass.

resolution The amount of detail that can be seen in an *object* or the *magnified* image of that object.

sample A small amount of something, such as water or blood.

sediment Tiny solid *particles* of mud, rock and animal or plant remains that settle on the bottom of rivers, lakes and seas.

solution A solid, liquid or gas which is dissolved in a liquid.

specimen An example of a particular thing that you want to look at with your microscope.

stage The part of a microscope where the *object* is placed. There is a hole in the middle of the stage so that light can shine through to illuminate an object from below.

stage micrometer A piece of clear film or a glass slide which has a measurement scale marked on it. It is placed on the *stage* of a microscope, like any other slide. A stage micrometer is often used with an *eyepiece graticule* to measure an *object*.

stereomicroscope A type of *optical microscope* which lets you see an *object* in three dimensions.

superbugs Types of *bacteria* that have become used to most *antibiotics* and are no longer destroyed or harmed by them.

transparent See-through. (An *object* that lets all the light pass through it is transparent.)

ultramicrotome A tool used for cutting extremely thin sections of a specimen for viewing with an *electron microscope*. (See also *microtome*.)

vaccine A medicine used to prevent certain diseases, such as measles or mumps.

vacuum A completely empty space, which contains no air or other gases. The tube of an *electron microscope* contains a vacuum.

virus A minute *organism* that can only reproduce inside the *cells* of other living things. Many viruses cause disease.

INDEX

Useful addresses

If you decide to become more involved in microscopy you may wish to contact a specialist organization. Below are some postal addresses that you might find helpful. For links to their Websites, go to the Usborne Quicklinks Website at **www.usborne-quicklinks.com** and enter the keyword "microscope".

Australia
Australian Microscopy and Microanalysis Society
The University of Sydney
Madsen Building, F09, Sydney
NSW 2006

Canada
Microscopical Society of Canada
2918 13th Avenue South
Lethbridge AL T1K 0T2

UK
The Quekett Microscopical Club
Subscription Manager
90, The Fairway
South Ruislip
Middlesex HA4 0SQ

The Royal Microscopy Society
37/38 St Clements
Oxford OX4 1AJ

USA
Microscopy Society of America
MSA Business Office
Suite 400, 230 East Ohio Street
Chicago IL 60611-3265

Microscopy suppliers

Go to Usborne Quicklinks for links to the websites of microscopy suppliers in the UK, USA, Canada and Australia
www.usborne-quicklinks.com

Other general websites

Here are some more fun and interesting microscopy websites. For links to these sites, go to the Usborne Quicklinks Website at **www.usborne-quicklinks.com** and enter the keyword "microscope".

• See objects close up using virtual microscopes.

• Go on exciting microscopic adventures.

• See fantastic close-up pictures of the natural world.

Websites

Acknowledgements

The Publishers would like to thank the following for their help and advice: DuPont; Histological Equipment Ltd; Harry Kenward, Environmental Archaeology Unit, University of York; Eric Marson, Northern Biological Supplies Ltd; Philip Harris Education; Alan Pipe and Jane Siddell, Museum of London Archaeology Service; Dr Rob Scaife, Department of Geography, University of Southampton.

Photo credits:

Key: top – t; middle – m; bottom – b; left – l; right – r; Science Photo Library – SPL.

Cover: Matthias Kulka/Corbis *(main)*; Visuals Unlimited/Corbis *(background)* **1:** David Scharf/SPL. **2-3:** Eye of Science/SPL. **4-5:** SPL. **6:** David Scharf/SPL *(b, m)*; Jan Hinsch/SPL *(t)*. **7:** Dr Kari Lounatmaa/SPL *(l)*; David Scharf/SPL *(t)*; Philippe Plailly/SPL *(b)*. **8:** Biophoto Associates/SPL *(t)*; James King-Holmes/SPL *(l)*; Eric Grave/SPL *(m)*. **9:** Andrew Syred/SPL *(t)*; Secchi-Lecaque/CNRI/SPL *(m)*; David Scharf/SPL *(r)*. **10:** Astrid and Hanns-Frieder Michler/SPL. **11:** Andrew Syred/SPL *(tr, mr, br)*. **12:** Astrid and Hanns-Frieder Michler/SPL *(l)*; David Scharf/SPL *(m)*; Dr Jeremy Burgess/SPL *(t)*. **13:** SPL *(l)*; SPL *(tr)*; Bruce Iverson/SPL *(r)*. **14:** Andrew Syred/SPL *(l, mr)*; Dr Jeremy Burgess/SPL *(m)*. **14-15:** SPL. **15:** SPL *(m, r, b)*. **16:** SPL *(l, m)*. **16-17:** SPL. **17:** Eye of Science/SPL *(m)*. **18:** Jan Hinsch/SPL *(l)*; Museum of London Archaeology Service *(t, br)*. **19:** Dr Rob Scaife, Department of Geography, University of Southampton *(l)*; Museum of London Archaeology Service *(m, b)*; Environmental Archaeology Unit, University of York *(tr)*. **20:** Dr Jeremy Burges/SPL. **21:** J.C. Revy/SPL. **22:** John Burbidge/SPL *(bl)*; David Scharf/SPL *(mr)*. **22-23:** Manfred Kage/SPL. **23:** BSIP VEM/SPL *(bl)*; Prof. P.M. Motta/Dept. of Anatomy/ University "La Sapienza", Rome/SPL *(tr, br)*. **24:** Prof. P.M. Motta/Dept. of Anatomy/University "La Sapienza", Rome/SPL *(l)*; Eddy Gray/SPL *(t)*; Philippe Plailly/SPL *(br)*. **25:** Carlos Munoz-Yague/Eurelios/SPL *(l)*; Richard Wehr/Custom Medical Stock Photo/SPL *(br)*; Hossler/Custom Medical Photo/SPL *(tr)*. **26:** Biophoto Associates/SPL *(l)*; K.R. Porter/SPL *(t)*. **27:** Bruce Iverson/SPL *(l)*; National Cancer Institute/SPL *(r)*. **28:** ©Carolina Biological Supply Co./Oxford Scientific Films *(l)*; Professors P. M. Motta and T. Naguro/SPL *(b)*. **28-29:** Professors P.M. Motta, S. Makabe and T. Naguro/SPL. **29:** Professors P.M. Motta and T. Naguro/SPL *(b, r)*. **30:** Professors P.M. Motta, S. Makabe and T. Naguro/SPL *(tl)*; Professors P.M. Motta and T. Naguro/SPL *(ml)*; Biophoto Associates/SPL *(b)*. **30-31:** Lawrence Berkley Laboratory/SPL. **31:** Francis Leroy, Biocosmos/SPL *(tr)*. **32:** Alfred Pasieka/SPL *(l)*; BSIP, VEM/SPL *(m)*; A.B. Dowsett/SPL *(t)*. **33:** CNRI/SPL *(l)*; Dr Kari Lounatmaa/SPL *(bl)*; Alfred Pasieka/SPL *(r)*. **34:** NIBSC/SPL *(b)*; Dr Kari Lounatmaa/SPL *(t)*. **35:** NIBSC/SPL *(m)*; NIBSC/SPL *(b)*; Richard J. Green/SPL *(t)*. **36:** Geoff Tompkinson/SPL *(l)*; Nancy Kedersha/Immunogen/SPL *(t)*. **37:** Dr Kari Lounatmaa/SPL *(bl)*; Biology Media/SPL *(br)*; Alfred Pasieka/SPL *(t)*. **38:** SIO/SPL. **39:** David Scharf/SPL. **40:** Sidney Moulds/SPL *(m)*; J.C. Revy/SPL *(t)*. **41:** Dr Jeremy Burgess/SPL *(l, m)*; Andrew Syred/SPL *(r)*. **42:** ©Dr B. Booth, GeoScience Features Picture Library *(l)*; Claude Nuridsany and Marie Perennou/SPL *(m)*; J.C. Revy/SPL *(br)*. **42-43:** ©Scott Camazine/ George Hudler/Oxford Scientific Films. **43:** J.C. Revy/SPL *(r)*. **44:** Dr Jeremy Burgess/SPL *(ml)*; Volker Steger, Peter Arnold Inc./SPL *(t)*; Andrew Syred/SPL *(b)*; David Scharf/SPL *(mr)*. **45:** Andrew Syred/SPL *(l, b)*; David Scharf/SPL *(m)*;

Dr Jeremy Burgess/SPL *(mr)*. **46:** M.I. Walker/SPL *(l)*; Manfred Kage/SPL *(tr)*; Jan Hinsch/SPL *(tl, br)*. **47:** Sinclair Stammers/SPL *(m)*; Alfred Pasieka/SPL *(b)*. **48:** G.F. Gennaro/SPL. **49:** Andrew McClenaghan/SPL *(l)*; Dr Jeremy Burgess/SPL *(r)*. **50:** Dr Jeremy Burgess/SPL *(tl)*; CNRI/SPL *(bl)*; BSIP VEM/SPL *(t)*; A.B. Dowsett/SPL *(m)*. **51:** David Scharf/SPL. **52:** Manfred Kage/SPL *(l)*; David Scharf/SPL *(tr)*. **53:** Dr Tony Brain/SPL. **54:** CNRI/SPL *(tl)*; David Scharf/SPL *(ml)*; ©Manfred Kage/Oxford Scientific Films *(bl)*; Biophoto Associates/SPL *(t)*; J.C. Revy/SPL *(br)*. **55:** ©Kjell B. Sandved/ Oxford Scientific Films *(bl)*; Claude Nuridsany and Marie Perennou/SPL *(br)*. **56:** David Scharf/SPL *(l)*; Claude Nuridsany and Marie Perennou/SPL *(m)*; John Walsh/SPL *(t)*. **57:** Biophoto Associates/SPL *(tl, l2, l3, bl)*; ©Harold Taylor Abipp/Oxford Scientific Films *(m)*; John Burbidge/SPL *(br)*. **58:** David Scharf/SPL *(l)*; K.H. Kjeldsen/SPL *(r)*. **59:** Dr Jeremy Burgess/SPL *(bl)*; David Scharf/SPL *(tm, br)*; Andrew Syred/SPL *(mr)*. **60:** Eye of Science/SPL *(b)*; Professors P.M. Motta and F.M. Magliocca/SPL *(t)*. **61:** K.H. Kjeldsen/SPL *(ml)*; Eye of Science/SPL *(tr)*; K.H. Kjeldsen/SPL *(b)*. **62:** Andrew Syred/SPL *(l)*; Manfred Kage/SPL *(m)*; Biophoto Associates/SPL *(r)*. **63:** ©Sinclair Stammers/Oxford Scientific Films. **64-65:** SPL. **65:** Mike McNamee/SPL *(ml)*; Alfred Pasieka/SPL *(b)*. **66:** Alfred Pasieka/SPL *(m)*; Dr Ann Smith/SPL *(tl)*; Dr Tony Brain/SPL *(tr)*. **67:** Dr Ann Smith/SPL *(tl)*; Philippe Plailly/Eurelios/SPL *(m, b)*; NASA/SPL *(r)*. **68:** Andrew Syred/SPL *(l)*; Manfred Kage/SPL *(r)*. **69:** Alfred Pasieka/SPL. **70:** Michael W. Davidson/SPL *(tl)*; Dr Clive Kocher/SPL *(bl)*; Sinclair Stammers/SPL *(r)*. **71:** John Walsh/SPL *(l)*; Michael W. Davidson/SPL *(r)*. **72:** G. Muller, Struers GMBH/SPL *(l, rm)*. **72-73:** G. Muller, Struers GMBH/SPL. **73:** Manfred Kage/SPL *(bl)*; Astrid and Hanns-Frieder Michler/SPL *(r)*. **74:** David Parker/SPL *(l)*; David Scharf/SPL *(m, t)*. **74-75:** George Bernard/SPL. **75:** Sandia National Laboratories/SPL *(l)*; Photo Courtesy of Paul McWhorter, Sandia National Laboratories *(tr)*; Toyota (GB) Ltd. *(mr)*. **76:** David Scharf *(t)*; Peter Menzel/SPL *(m)*; IBM Research Division, Zurich *(b)*. **77:** IBM Corporation, Research Division, Almaden Research Center *(bl)*; IBM Research Division, Zurich *(tl, tr.)*; Professor Harold G. Craighead, Cornell University, New York *(br)*. **78:** Julian Baum/SPL. **79:** Science Museum/Science and Society Picture Library. **80:** Dr Jeremy Burgess/SPL *(l, tm)*; SPL *(bm)*; Science Museum/ Science and Society Picture Library *(r)*. **81:** Science Museum/Science and Society Picture Library *(tm)*; By permission of the President and Council of the Royal Society, London *(br)*. **82:** SPL *(l)*; ©The British Museum *(m)*; Jean-Loup Charmet/SPL *(r)*. **83:** Custom Medical Stock Photo/SPL *(l)*; D. McMullan/SPL *(t)*. **84:** IBM Research Division, Zurich *(l)*; P. Plailly/Eurelios/SPL *(r)*. **85:** David Parker/SPL. **86-87:** Philip Harris Education. **88-89:** Howard Allman. **90:** Philip Harris Education.

Cover designer: Stephen Moncrieff   Photographic manipulation: John Russell
Every effort has been made to trace and acknowledge ownership of copyright. The publishers will be glad to make suitable arrangements with any copyright holder whom it has not been possible to contact.

Printed in Dubai